MAKING WATER SECURITY

A MORPHOLOGICAL ACCOUNT OF NILE RIVER DEVELOPMENT

Hermen Smit

MAKING WATER SECURITY

A MORPHOLOGICAL ACCOUNT OF NILE RIVER DEVELOPMENT


DISSERTATION


Submitted in fulfillment of the requirements of

the Board for Doctorates of Delft University of Technology

and

of the Academic Board of IHE Delft Institute for Water Education

for

the Degree of DOCTOR

to be defended in public on

Tuesday 17 December 2019 at 15.00 hours

in Delft, the Netherlands


by


Hermen SMIT


Master of Science in Civil Engineering, Delft University of Technology

Master of Science in Environment and Resource Management,
Vrije Universiteit Amsterdam

born in Amsterdam, the Netherlands

This dissertation has been approved by the promotor and co-promotor

Composition of the doctoral committee:

| | |
|---|---|
| Rector Magnificus TU Delft | Chairman |
| Rector IHE Delft | Vice-Chairman |
| Prof. dr. ir. P. van der Zaag | IHE Delft/ TU Delft, promotor |
| Dr. ir. R. Ahlers | SOMO, co-promotor |

Independent members:

| | |
|---|---|
| Prof. dr. A. El-Battahani | University of Khartoum, Sudan |
| Dr. F. Molle | Institut de Recherche pour le Développement, France |
| Prof. dr. E.B. Zoomers | Utrecht University |
| Prof. dr. mr. ir. N. Doorn | TU Delft |
| Prof. dr. ir. S.N. Jonkman | TU Delft, reserve member |

This research was conducted under the auspices of the SENSE Research School for Socio-Economic and Natural Sciences of the Environment

Smit, Hermen (2019) Making Water Security: A Morphological Account of Nile River Development. Leiden: CRC Press/Balkema

CRC Press/Balkema is an imprint of the Taylor & Francis Group, an informa business

Published by:
CRC Press/Balkema
Schipholweg 107C, 2316 XC, Leiden, the Netherlands
Pub.NL@taylorandfrancis.com
www.crcpress.com – www.taylorandfrancis.com
ISBN: 978-0-367-46004-4

# Contents

# Contents

# Acknowledgements

This morphological account of Nile river development is a product of the work of many. For me, the work on this project started in 2008, when applying for a project on "institutionalizing compensation flows…for improved [Nile] river basin management" (UNESCO-IHE et al. 2007, p5). Today my commitment to changing Nile river management remains, but I found no single way of doing this by institutionalising compensation flows. Rather I have learned about how and why the idea of 'compensation flows' is itself a product of modern river improvement. This insight was triggered by multiple accounts of redistribution of land, water and soil that I could not grasp in terms of compensation. Trying to understand how these accounts relate, forced me to learn to appreciate the ambiguities of securitizing the river, rather than reducing them to fit a fixed framework of water security. This has been a confusing and extremely rewarding endeavor, during which I have met and learned from many wonderful people. I am grateful to all who contributed to this project, and therewith shaped me as a person, researcher and teacher.

First, I would like to thank the many people who taught me about the politics of irrigation and drainage in Choke, Gezira and Waha. In particular, I thank the Eljeli family in the Gezira, Itataku and Yegirmal and the Michael Idder in Choke and all at the Hydraulics Research Center in Wad Medani for making me feel part of their families and for sharing with me their joy, sadness and knowledge. Special thanks also to my closest collaborators in Ethiopia and Sudan for your friendship and comradeship: Rahel Muche, Tesfaye Muluneh and Tefera Goshu for an incredible year in the Choke Mountains. Yasir Salih, Abdel Nasr Khidr, Abu Obieda Babiker, Mohaned, Khalid Hajelneel, ElFadil Abd Elrahman, Elfatih, Siddig Yousif, Mohammed Shazera, Eshraga Osman, Yasir Mohamed, Sami El-Tom and Emma Aalbers for the two irrigation seasons during which we dug into the Gezira canals together. Thanks to Nagwa, Abdallah, Murtada, Safia Mohamed, Blaze Horn, Mazza Dolib, Nutaila El Moghraby, Khalid Biro, Edwin Rap for your engagements in the making of multiple irrigation realities at Waha. Thanks also to all who helped me in the libraries of IHE Delft, TU Delft, the ministry of Water Resources and Irrigation in Egypt, the ministry of Water and Energy in Ethiopia, the Hydraulic Research Center in Sudan and the Gezira scheme in Sudan.

I would also like to thank the students and colleagues at IHE Delft who together made a great home for digesting what I learned in Ethiopia and Sudan, and for learning more about how water science and education shape the Nile River. Pieter van der Zaag has been the creative inspirator of this Nile project. Your way of analysing how the development, appropriation and uses of technologies shape struggles over land, water and infrastructure have profoundly influenced my way of looking at water development. Thank you for your friendship, energy and faith in the project, which were key in getting it done. Rhodante Ahlers' engaged scholarship and critical activism have inspired me throughout the research. You pushed the theoretical and methodological boundaries of the research. Thank you for sharing your intellectual creativity, for introducing me to geography, and for reading and commenting on so many drafts. The times we spent together in Delft, Amsterdam, Seattle and Choke have made me a better researcher and teacher. The MSc students who worked with me on their thesis projects about practices of irrigation and drainage development have also shaped my thinking about the materiality of making water security. In particular, I want to thank Atsbha Bhrane, Aytenew Tatek, Kidist Bekele, Amsalu Kussia Chamaysha, Sachin Tiwale, Cristiano Von Steinkirch de Oliveira, Koen Mathot, Ertiban Woldegebriel, Mathilde Köck, and Akewak Gobosho.

The Blue Nile Hydrosolidarity Project (by Addis Ababa University, University of Khartoum, IWMI, Vrije Universiteit Amsterdam and UNESCO-IHE) and the Accounting for Nile Waters Project (Hydraulics Research Center Sudan, IHE Delft, IWMI, Gender center for Training and Research Sudan, RISE - American University in Cairo) under which this research was implemented, provided exciting workshops and seminars for delving into the ambiguities of Nile science for development. The 'in search of hydrosolidarity' seminars with Yasir Salih, Rahel Muche, Ermias Teferi, Sirak Tekleab, Abonesh Tesfaye, Melesse Temesgen, Pieter van der Zaag, Stefan Uhlenbrook, Rhodante Ahlers, Yasir Mohamed, Jochen Wenninger, Wubalem Fekade, Solomon Abate, Roy Brouwer, Yacob Arsano, Atta El Battahani, Eshraga Osman, Tesfay Gebremichael, Belay Simane, Workneh Nigatu, Huub Savenije and Seifildin Abdalla have influenced my thinking about 'Nile development' in productive ways. I further thank the Blue Nile Solidarity group for the many helpful comments I received. Thank you to Tesfaye Muluneh, Safia Mohamed, Atta El-Battahani, Khalid Biro, Yasir Mohamed, Wim Bastiaanssen, Tina Jaskolski, Hassan Husseiny, Livia Peiser, Edwin Rap, Emanuele Fantini and Margreet Zwarteveen of the Accounting for Nile Waters project for many inspiring debates we had about the making of multiple realities of water accounting. Special thanks to Edwin Rap for the exciting collaboration in Sudan and for sharing with me your approach to analyzing and documenting practices of water development; to Emanuele Fantini, for sharing your creativity in pushing new collaborations over Nile waters; to Elien Phernambucq for your help with the Nilewaterlab.org, which presents much of Chapters 2-5 of this dissertation in a more visual form; to Atta El-Battahani for generously sharing your knowledge about Sudanese politics and for being an example in 'science for action'; to Khalid Biro, for teaching me about practices of 'water accounting'; and to Margreet Zwarteveen for your close readings of draft chapters and especially for inspiring conversations about what it means and takes to break the boundaries between water science and politics - I hope these will continue for a long time.

Many people outside the above mentioned working groups and organisations provided energy and ideas for this project. Special thanks to Alex Bolding for the inspiring times we shared in Arba Minch and Frederiksoord. Judith Kaspersma, Dieneke Smit, Lieke Oldenhof and Ceren Sezer: thank you for the good times we had reading and writing together in Lettelbert, Woudrichem, Buurse and in the TU Delft library. Thank you Gavin Bridge, René Lefort, Alula Pankhurst, Andres Verzijl, Jonas Wanvoeke, Saskia van der Kooij and Marcel Kuper for providing comments on draft chapters. And thank you Arjen Zegwaard for sharing your fascinating ideas about muddy understandings of deltas.

The people working on the MSc programme in Water management and Governance, the Graduate School and in the Water Governance and Water Management groups at IHE Delft provided me - and continue to provide me - with a working environment that is warm, never dull and often inspiring. Thank you for that Michelle Kooy, Jaap Evers, Bosman Batubara, Adriano Biza, Akosua Boakye-Ansah, Andrés Cabrera Flamini, Gabriela Cuadrado-Quesada, Emanuele Fantini, Joyeeta Gupta, Shahnoor Hasan, Rozemarijn ter Horst, Hameed Jamali, Frank Jaspers, Klaas Schwartz, Zaki Shubber, Phil Torio, Mireia Tutusaus, Anna Wesselink, Angela Bayona Valderrama, Cristóbal Bonelli, Natalia Reyes Tejeda, Irene Leonardelli, Susanne Schmeier, Jenniver Sehring, Tatiana Acevedo Guerrero, Jeltsje-Kemerink Seyoum, Emanuele Fantini, Margreet Zwarteveen, Emilie Broek, Patricia Darvis, Zeliha Bağci, Nora van Cauwenbergh, Yong Jiang, Ilyas Masih, Janez Susnik, Marloes Mul, Jonatan Godinez Madrigal, Wim Douven, Angeles Mendoza, Leon Hermans, Abebe Chukalla, Bert Coerver, Claire Michailovsky, Elga Salvadore, Bich Tran, Fikadu Fetene Worku, Solomon Seyoum, Sara Masia, Pieter van der Zaag, Susan Graas, Jetze Heun, Maria Rusca, Rhodante Ahlers, Yasir Mohamed, Jaqueline Koot, Selda Akbal, Jos Bult, Gerda de Gijsel, Lilian Hellemons, Ineke Kleemans, Yared Abebe, Patricia Trambauer, Veronica Minaya Maldonado, Aline Saraiva Okello, Hans Komakech, Caroline de Ruiter, Erwin Ploeger, Marlies Baburek, Joyce Membre, Dwi Soewanto, Ben Stuijfzand, Schalk Jan van Andel, Charlotte de Fraiture, Jolanda Boots, Floor Felix, Paula Derkse and Anique Karsten. Special thanks to Tatiana Acevedo Guerrero, Jeltsje Kemerink-Seyoum, Emanuele Fantini and Margreet Zwarteveen for friendship and inspiration while teaching together. To get the writing done

I ran away from you all a couple of times: thanks to the Health Care Governance group at Erasmus University for providing a warm shelter on these occasions.

I am grateful to IHE Delft, Netherlands Organization for Scientific Research (NWO-WOTRO) and the DGIS UNESCO-IHE Programmatic Cooperation (DUPC) for financially supporting this research as part of the project "In search of sustainable catchments and basinwide solidarities in the Blue Nile River Basin". I am grateful to CGIAR's Nile Water Land and Ecosystems programme for financially supporting the research for chapter 5 as part of the project "Accounting for Nile Waters- Connecting Investments in Large Scale Irrigation to Gendered Reallocations of Water and Labor in the Eastern Nile Basin."

My family and friends outside academia supported me all along the project: thank you Eri, Henk, Annet, Peter, Bram, Cynthia, Dieneke, Thijs, Maurits, Peter, Wim, Jory, Hanneke, Alex, Frank, Natalia, Martijn, Ingrid, for your love, inspiration and encouragement. I am most grateful to Lieke for living with me and supporting both myself and the research. Your love, ideas and ideals have become a big part of this project. I thank Juul and Rokus for making me complete it.

**1**

# INTRODUCTION

## 1.1 Introduction

What makes water security on the Nile? Far away from the river, in water ministries, universities and international development institutions, water security is often talked about as a matter of ensuring the "availability of an acceptable quantity and quality of water... coupled with an acceptable level of water related risks to people, environment and economies" (Grey and Sadoff 2007: p545, GWP 2014a, UN Water 2013). The Nile has become an emblematic case to illustrate and underscore the need to intensify technical, governance and diplomatic efforts to adapt to increasing climate uncertainties and fend off resource conflicts (see e.g. Homer-Dixon 1994, Wolf 1999, Subramanian et al. 2012, Link et al. 2016). From 2000 onwards, this water security storyline has served as the justification for an unprecedented number of new dam and irrigation projects along the Nile. With hydropower development presented as key in the fight against climate change and irrigation development as essential to meet growing global food requirements, governments collaborate with companies around the world to finance and construct new projects. Yet, by accounting for these projects in terms of kilo-watt-hours of electricity and tons of food produced, important questions of how these projects change the river and to whose benefit remain unanswered.

This dissertation approaches water security in a different way. Rather than understanding it as the availabilities and reliabilities of water for anonymous and abstract 'people, environments and economies', it focuses on the changing course of the river. The course of the river can reveal the contested relations that are solidified in its bed, that drive the institutions and shape the infrastructure of the river. Understood in this way, the specific meanings enforced by new projects of water security become apparent.

The changing morphology of the river – its form and structure – serve as an empirical and theoretical device to understand how infrastructures and discourses of Nile water security change with the river they redistribute. The sediments that shape the river bed are an obvious starting point here. They facilitate flows of water to some, while blocking access to others. This is not only about movement but also about fixity. And it is as much about material flows as it is about discourses of development. Fixing water, and stereotypes of water users, engineers and leaders in large dams and irrigation bureaucracies along the river facilitated their mobilization for colonial production[1]. But fixing sediments in irrigation canals and behind dams also posed challenges to these very projects. Clogged water infrastructures prompted Egyptian rulers of the 19th century to conscript labour to

---

[1] The point of dams 'fixing water to increase its mobility' comes from Ahlers (2005).

clear canals, and engineers to design dams in such a way that they could be emptied of sediments during the flood season (Hamdan 1972). It is these colonial dams, and the hierarchies that were institutionalised with them, that influence new movements of water, sediment, people, food, money and science that rearrange the river today.

Examining these interactions enables an understanding of how the forms, roles and possibilities of hydraulic infrastructures on the Nile changed over time. It shows how large hydraulic infrastructures and institutions throw long spatial and temporal shadows over the river. Taking the sediments of the Nile as the starting point enables a relational definition of water security, one that acknowledges that rivers are a product of earlier relations sedimented in the river bed and that also appreciates the possibilities of millions along the Nile working continuously to remake these relations. These relations consist of collaborations and interactions between people, infrastructures, silt and water that evolve as part of highly unequal but undetermined struggles over water, land, labour and power.

The central argument of this dissertation is that indicators, institutional principles or infrastructural designs for water security are never independent from the rivers they seek to measure, govern or develop. The knowledge and the interventions that these theories of water security inform are always more than representational: they also help bring into being new morphologies, new distributions, new patterns of water security. Understanding how water users, engineers and researchers produce power laden patterns of water distribution through superimposing river basin development projects is central to this research, as these relations re-define the responsibilities of those involved in these projects.

Trained as a water engineer in the heartland of Dutch water science, I am expected to confidently engage in one of Holland's most famous export products: water management. Yet, during my first job as a consultant in Bangladesh a large chunk of this confidence eroded with the dikes of the rivers I worked on maintaining. It was bad enough that the land protected by the embankments designed by my predecessors stopped accreting and subsequently became vulnerable to flooding. What was as troublesome was that the divisions that shaped the lack of public participation in maintenance we sought to overcome had been inscribed into the landscape by decades of projects ordered by the water development board that I worked for. This sparked my interest in my own role and responsibility in working on shaping rivers. Was it a coincidence that the engineers who were transferred from India to 'develop' Egypt in the late 19th century were all male white engineers, like me? How was the project I engaged in different from theirs? Taking up a PhD fellowship in a project that explored the willingness of Nile water users "to invest in sustainable practices and catchment-wide solidarities" (UNESCO-IHE et al. 2007, p9) provided the space to explore these questions.

Over the 20th century the Nile has changed tremendously. The volume of water flowing to the sea was decimated (Sutcliffe and Parks 1999). Dams are no longer constructed to store part of the flood waters and pass the sediments; they now store entire Nile floods. With the increase in cultivation of the Ethiopian highlands, the sediment load of the Blue Nile also increased fivefold (Chapter 3), posing major problems of fertility decline in the highlands and of siltation of dams and irrigation schemes more downstream. The efforts of our research team to substantiate the link between erosion in Ethiopia and sedimentation in Sudan aligned with ideas of governments, companies and development organisations to institutionalise basin wide solidarities through new projects of hydraulic development, benefit sharing and payments for environmental services.

Yet during the two drainage seasons we spent in the Choke Mountains and the two irrigation seasons in Sudan's Gezira irrigation scheme the problems of soil erosion and sedimentation far exceeded the lack of 'participation' and 'willingness to pay' for new hydraulic investments in water security. While discussing about how silt particles flow down the Choke Mountains in Ethiopia and settle in irrigation canals of the Gezira scheme in Sudan, I learned about how the superimposition of centuries of struggles over drainage and irrigation shaped class, age and race relations along the river. Appeals to water security tend to obscure, although with decreasing success, the making of these relations. Indeed, in media, bureaucratic and academic reports of Nile development, water security is increasingly framed as a matter of growing returns to water services and compensating those who supposedly lack sufficient resilience.

Nevertheless, I cannot and do not want to separate myself from the modern scientific logic that informs such frameworks. Not only is this research forged by my acquaintance with this logic during my training as a civil engineer at Delft University of Technology. The work of my fellow researchers in the Blue Nile Hydrosolidarity project reminded me that its tools can be powerful in revealing both unexpected patterns of distribution of Nile water and silt, as well as collaborations between water users and researchers from Sudan and Ethiopia that few had imagined possible[2]. The purpose here, is to understand how the highlighting of patterns and the making of collaboration is not 'just technical' but always at the same time constituted by and constitutive of Nile politics. In other words, I am

---

[2] For reports of some of these unexpected patterns of water and sediment and collaborations, see Temesgen et al. 2012, Gebremichael et al. 2013, Tesfaye 2013, Salih 2014, Tekleab 2015, Teferi 2015, Osman 2015 and Kahsay 2017.

interested in how scientific work, like other work, is always informed by and pushing particular visions of living with water that benefit some and not others.

My aim in this dissertation is to analyse how scientists, engineers and water users engage in rearranging the morphology of the Nile and thereby shape their relative positions vis à vis each other and the river. Thus, it intends to contribute to the forging of new alliances to support more equitable and sustainable forms of Nile development. This chapter introduces the concepts and methodology that inspired and form this dissertation. 1.2 provides a brief genealogy of the modern concept of water security. 1.3 introduces the concepts that inform the morphological approach to analyse and displace the modern separations between technologies and politics through which modern water security comes in to being. 1.4 sets out the structure of the morphological account of Eastern Nile development presented in this dissertation.

## 1.2 The hydraulic mission and the rise of water security as a modern concept

Water security as a concept gained popularity in academic literature around the turn of the millennium (Cook and Bakker 2012, Clement 2013). Yet I would like to trace the roots of the concept to what came to be known as the 'hydraulic mission' of modern river basin development in the 19th century (Molle 2009). By that time, the looting of foreign territories itself was no longer a central motive of imperial expansion. With wheat, rice, cotton and sugar becoming global commodities, the availability of fresh water to grow these crops became a key condition for imperial industrialization. The 19th century and early 20th century did not only see a surge of large hydraulic construction in the arid plains of the western US (Worster 1985) and Southern Europe (Swyngedouw 2015). The securitization of water for sugar and cotton production also justified colonial expansion around the tropical world, e.g. in Indonesia, Latin America, India and North Africa. This was the context in which William Willcocks - one of Britain's famous irrigation engineers trained in India and then chief engineer of the Egyptian ministry of public works – claimed that the cultivation of cotton in 19th century Egypt had made "the securing of an abundant supply of water all the year round ... the problem of the day" (Willcocks 1904, p73).

The modern scientific separation of nature and society, with the former serving the development of the latter, was central in narratives that justified colonial expansion (Escobar 1995). Mathematics, hydrology and hydraulics were not only instrumental to construct new large engineering works (Molle et al. 2009), but the meticulous scientific documentation of the Nile was also used to downplay violence and glorify the superiority of the knowledge of European engineers (Said 1978). Envisaging the 'taming' of the river, Winston Churchill imagined in 1898 how Garstin's plans for irrigation along the Nile

> *"may in their turn prove but the preliminaries of even mightier schemes until at last nearly every drop of water which drains into the whole valley of the Nile, preserved from evaporation or discharge, shall be equally and amicably divided among the river-people – and the Nile itself, flowing for three thousand miles through smiling countries, shall perish gloriously and never reach the sea"* (Churchill 1899, p411).

Colonial water security was thus portrayed as a civilising project for the development and betterment of mankind. Like water, people, especially non-European people, were rendered as vital resources whose development would be secured by giving them a place in imperial production.

Irrigated cotton and sugar estates, erected in the name of civilization, would not only fuel but also produce the limits to these empires. The widespread foreclosures of farms by farmers who lacked the capital for intensive cultivation of market crops created a large group of landless peasants that posed a continuous threat to large capitalist farmers (Kautsky 1988 [1899]). After an initial exponential growth of cotton and sugar production in Egypt, the accumulation of salts carried with irrigation water led to diminishing returns and the abandoning of some of the lands converted for imperial agriculture. With growing revolts, imperial production increasingly in crisis and the costs of war rising, the British Empire imploded in the first half of the 20th century (Tomich 1994, Goswami 2011, El Shakry 2007, Tvedt 2004).

After the Second World War, European investors and local estate or plantation holders were desperate to protect their assets in Africa, Asia and Latin America. These investors worked with the newly emerging global superpowers seeking to expand their markets for grain and territorial control, to secure water and power through 'national' industrial development (Ekbladh 2002). In the process new markets were opened up for wheat and fertilizer and cheap supplies of tropical products continued to flow into the USSR and USA. Mega-dams to provide high input modern agriculture became key political instruments of the Cold War era (Waterbury 1979, Howell and Allan 1994). By the 1960s, Fordist and Khrushchevian models to finance and construct state controlled dams in irrigation schemes were thriving around the postcolonial world (Sneddon 2012, Kalinovsky 2013). The revolution in yields brought about by these schemes proved to be costly: They would be mortgaged on cheap Southern labour, growing public debt, depleted rivers and an addiction to western fertilizers and wheat for millions no longer producing food themselves. And when a new financial crisis emerged in the 1970s and debts were due, Southern consumers of wheat were made to bear the brunt of these costs by structural adjustment (McMichael 1997, Moore 2015).

The World Bank suspended the financing of large state-owned infrastructure and instead launched institutional reforms. Meanwhile US food aid was replaced with food trade at market prices. With IMF debt relief conditioned on the devaluation of currencies to reduce production costs, this led to sharply reducing real wages in the postcolonial world. The 'market' proved even more expensive for people in former colonies when social scientists inspired by Malthusian theorizations of the commons linked the increasing demand for environmental resources by a growing population to an increasing risk of conflict (Floyd and Matthew 2013). Drawing on such frameworks, international organisations like the World Bank insisted to make water users pay for water to increase its efficient use (World Bank 1994). By rendering environmental degradation as a global problem, waters long allocated were opened up for redistribution in global market terms.

This was the context in which water security crystallized as a popular analytical concept used today by governments, companies and international organisations. Its strength lies in its welding together of two discourses that are partly contradictory. First, the market is mobilized in response to depleting rivers and the concerns of rising environmentalist movements of the 1970s. Environmental problems could be solved by valuing the environmental costs that had mistakenly been left unaccounted for until then. Once mechanisms are in place to internalize these costs, is the now widespread mantra amongst development organisations and government officials, the market will assign water to its most efficient users, including the environment. Second, with communism and capitalism no longer seen as major threats to global security, politicians and scientists labelled environmental resource scarcity as the single most important threat for human violent conflict (Wolf 2005). This legitimized the fast-tracking of new projects of water development in the name of security (Warner 2008, Cascão 2009a).

The joining of environmental security and military security agendas in a new agenda of 'market alarmism' informs and explains the wide range of water security frameworks produced by international policy actors such as the Global Water Partnership (2000, 2014, 2015), UN (FAO 2000, UN Water 2013), World Resources Institute (Gleick and Iceland 2018), and the World Bank (Subramanian et al. 2012). Despite the enormous diversity among these frameworks noted by scholars of water security (Zeitoun et al. 2016, Lankford et al. 2013, Pahl-Wostl et al. 2016, and Cook and Bakker 2012), the most widely used and cited water security frameworks share three broad characteristics:

1. A modern rationalisation and commodification of water as a complex adaptive system consisting of tradable ecosystem services forms the basis for analysing water distribution and development.

2. A replacement of probability based models for state led comprehensive river basin management by models that focus on the facilitation of individual water users/entrepreneurs to adapt to changing circumstances and prepare for an uncertain future (Bakker and Morinville 2013, Zegwaard 2016).

3. A sense of accountability for water development that is reduced to a separate audit-like account of economic performance, with the transparency and integrity of the individual entrepreneurs figuring prominently in proclamations of what 'good governance' is or should be (Rhodes 1996, Zwarteveen et al. 2017).

This water security agenda informs a wide range of projects by governments, international organisations, NGOs and other 'civil society' organisations. These are overwhelmingly framed in the language of market adaptation and technological, or governance, innovations. This agenda has also sparked intense debate. Not only do the market based frameworks for fostering water security sit uncomfortably with the 'fundamental human right to water' often pushed by the same organisations (Gupta et al. 2010). There is also mounting evidence that 'market based solutions' often work to erode the very adaptive capacities of large groups of people these projects claim to strengthen (Bakker 2010, Kaika 2017, Watts 2013, Warner et al. 2018).

In the discussion there has been strikingly little attention for how different groups of people and infrastructures, waters and soils along the river engage in shaping the processes through which securitization of the river takes form. Instead of providing yet another theory of water security that demarcates the 'safe operating subspace' by mapping 'physical resource availability, infrastructure and economic choices' to achieve the Sustainable Development Goals (Srinivasan et al. 2017, p12), I propose to examine how the manoeuvrable spaces and indicators to gauge them take form with the people who use them and their uses of the river. Such a historical and material analysis does not follow from combining, or integrating, social and natural or the discursive and material dimensions of water security. Rather, it turns the attention to how infrastructures, institutions and science of water security are the contested outcomes of struggles over hydraulic construction.

## 1.3 Morphologies of the Nile river bed, state power and water science

To examine how frameworks of water security are produced with and form part of contested water infrastructures, institutions and science, this dissertation analyses:

a.  how the expanding set of techniques through which populations of water users are conceived in terms of market and resilience (re)produces particular structures of control and domination;

b.  how the engagements of water, infrastructures, water users and scientists (however powerful or marginal) involved in securitizing the river shapes possibilities of river development in unexpected ways.

My aim is  to move the analyses of water security from  a prescriptive framework to inform the organisation of water distributions OR as an empirical description of particular manifestations of water distribution, governance or user capacities/knowledge, towards an analysis of the simultaneously changing forms AND the structures through which water security is understood.

I draw inspiration here from Goethe's 'morphological approach'[3] to study the (re)making of nature and modern science (Goethe 1987 [1776-1832]) and a variety of scholars who have since sought to displace the modern scientific separation between technology and politics[4]. Goethe came up with his approach in response to a range of challenges to the 18th century Linnaean ordering of taxa. The division of science into disciplines and the specialization of Linnaean botany facilitated a rapid expansion in the discovery of species. Yet with the discovery of more and more species, Linnaeus' system which classified taxa in 'natural categories' according to their 'God given' and 'obvious' features increasingly came under pressure. Not only had the categories of the classification to be revised several

---

[3] The summary of Goethe's morphological approach in this paragraph draws on Wellmon (2010).
[4] Accounts concerned with breaking of boundaries between the technical and the political that inspired me are: Haraway (1991), Latour (1993), Haraway (1997), Mitchell (2002), Raffles (2003), Tsing (2005), Ong (2006), Li (2007) Yeh (2012) and Moore (2015). I have been also inspired by the following accounts of scholars who have substantiated this debate in different ways in the water literature by analyzing relationships between the making of water technologies, institutions and science, and the management and governance of water: Mollinga and Bolding (1996), Mosse (1997), Swyngedouw (1999), Van der Zaag (2003), Bakker (2003), Bolding (2004), Kooy and Bakker (2008), Molle (2009), the special issue in Social Studies of Science edited by Barnes and Alatout (2012), Barnes (2014), Meehan (2014), Mollinga (2014), Akhter and Ormerod (2015), the special issue in Water Alternatives edited by Obertreis et al. (2016), Kemerink et al. (2016), Zwarteveen et al. (2017), Acevedo Guerrero (2018) and Batubara et al. (2018).

times. Species of which features changed a lot over their lifetime were also difficult to classify. The resulting crisis in taxonomy was the starting point for Goethe's fundamentally different approach to knowing plants. While he admired and used the efficiency of the Linnaean taxonomic system for determining particular species, he refused to accept the highly specialized knowledge generated as God given or essential. For him, increasing knowledge was not just a matter of the aggregation of discoveries. It was also about "how individuals can relate their own modes of inquiry to historical forms of knowledge" (Wellmon 2010, p154). Goethe's morphological approach was therefore not only directed to the features of plants but also to the mechanisms through which individuals changed the shifting frames through which plants were understood (Goethe 1987, 477). This required not just close empirical observation of plants but also a continuous reflection on and adjustment of the interpretative frame through which he – as a poet-scientist – himself developed. Morphology, for Goethe, thus would not only refer to the sediment particles that make the river bed, but also to the ways in which their ordering changes along with the users, scientists and engineers mapping and developing them. This brings to the fore how new scientific orderings of plants, water and sediments facilitate some modes of agriculture and not others. And it is through this relation of human involvement in the purposeful (re)ordering of nature that he recovered science itself as a deeply ethical practice. Goethe's morphological approach thus resists the urge to separate or choose between the empirical observation of a river bed and the rational order through which it is known. It instead turns to the relations between the subject (scientists) and the object (plants or sediments), making explicit the historical and ethical nature of the orderings through which they relate and transform together.

Some two centuries after Goethe's involvement in the crisis in botany, a new crisis of disciplines has erupted. The rising ecological and social inequalities that emerged along with projects of industrial development have put the modes of governance and science that enabled them increasingly under pressure. Not only do expanding complexities of global problems like climate change and rapid urbanisation pose challenges to universities and bureaucracies divided by disciplinary and sectoral boundaries. With the fall of the commercial and government bureaucracies they supported, also the roles and rules of experts themselves are increasingly questioned (Mitchell 2002, Esteva and Escobar 2017). In the face of climate change denial, scientists from a wide range of disciplines increasingly appreciate the need to make explicit the interests and values (or normative orders) enclosed in the facts produced if they want to uphold their scientific claims (Latour 2018). Goethe's observation that facts are never universally true, but always a product of the practices of their making is now as relevant as it was when he first made it (Feyerabend 1981, Latour 1987, 2018, Haraway 2016).[5]

Meanwhile, disciplines and knowledges have further specialized. The morphology concept has been taken up by a variety of disciplines. In one of these disciplines, hydraulics, mid-19th century river engineers took up the morphology of the river, i.e. the relation between the flow regimes of water and sediment in the river and the form of the river bed, as a major object of river engineering. Estimating the energy for geomorphic 'work' in a river by calculating shear stress created by particular flows of water through a river channel (Du Boys 1879), they analysed how the river changed through increasing the uptake of particles whenever the flow of water increased. As highlighted by Manning, one of the founding fathers of modern hydraulics, about the formulae he thus derived:

> "if modern formulae are empirical with scarcely an exception, and are not homogeneous, or even dimensional, then it is obvious that the truth of any such equation must altogether depend on that of the observations themselves, and it cannot in strictness be applied to a single case outside them" (Manning 1895, px cited by Liu 2014, p136).

And yet it was exactly by this wider application of formulae like Manning's through which the politics of Nile development unfolded. British engineers (like Egyptian state engineers and myself after them) certainly used Manning's formula to design canals in Egypt and quantify the Nile's water flow. Rather than simply applying the formula to concrete objects however, the practices through which the formula was adjusted and through which designs were adapted to deal with silty water and lands that were not level – often based on centuries of experience of Egyptian farmers - itself formed new Nile science. As highlighted by Mitchell this was not just a generative process. As the intricate morphology of flood irrigation in the Nile delta was reordered by large dams and punctuated by irrigation canals, hundred thousands of farmers lost part of their control over irrigation and with it saw part of their knowledge 'taken away' from them (Mitchell 2002, p37).

---

[5] Feyerabend linked explicitly to Goethe when he wrote that "Newton... did not give the explanation [of light] but simply re-described what he saw in order to turn it into a physically and useful phenomenon. And in this description he introduced the machinery of the very same theory he wanted to prove. Goethe's question 'for how should it be possible to hope for progress if what is inferred, guessed or merely believed to be the case can be put over on us as a fact?' addresses itself to this feature of [Newton's] theory" (Feyerabend 1981, p44).

Building on the above insights the remainder of this section develops a morphological approach to Nile development that appreciates the configuration of the river bed as a contested history of technological change. First, the concept of 'technopolitics of water security' is introduced to understand how modern techniques of calculation shape the securitization of the river. Thereafter, the morphology of the river is mobilized as a contested product of work to understand the materialization of differences along the river. Third, the framework elaborates on how the *technopolitics of classification* and the *materialization of differences in the river* are deeply intertwined and constitute each other. Focusing on how the river and the ways it is valued shape each other, the section on 'making water science and ethics' highlights what the implications are of this relation for the science of Nile development.

## Technopolitics of water security

To understand how the making of western science has become a powerful tool in rearranging Nile waters and populations, I build on Timothy Mitchell's and Jessica Barnes' analyses of the techno-politics of modern Egyptian development (Mitchell 2002, Barnes 2014). This concerns the continuous efforts of mapping, standardization and calculation by which business families, government agents, foreign investors and development organisations make the networks and projects through which they expand their control over land, water, labour, food and finance capital appear as universal signs of civilization (or progress or development). In this dissertation I trace the practices of water science, engineering and organisation through which frameworks of water security are *performed* (Butler 1990) as natural and neutral, i.e. as independent from the particular elements/histories that (in)scribed them.

Michel Foucault's work on *"Security, Territory, Population"* provides an important starting point here. In this work, Foucault traces the history of modern governance (or development) to the moment when *"eighteen century Western society took on board the fundamental biological fact that human beings are a species"* (Foucault 2007, p1). This rendering of populations made its relations to the environment amenable to procedures of standardization and normalization. Quite unlike sovereign rulers who determined what was permitted and what was not, this 'apparatus of security', as he called it, established an optimum "and a bandwidth of the acceptable that must not be exceeded" (ibid, p6). Extremes were no longer a priori valued as good or bad but viewed as part of *normal* distributions.

The wide array of techniques and micro-practices through which modern expertise is mobilized to order people's houses, canals, and judicial systems, rearrange what gets to be counted as 'developed' or 'legal'. This not just works to (self)discipline individuals in highly particular ways, but in the process, these techniques and practices assumed the very "appearance of abstract, nonmaterial forms", indeed suggesting the existence of, or bringing into being, structures like the state, the economy and society (Mitchell 1999, p77). The construction of modern dams and irrigation schemes not just took away land, water and knowledge from millions along the Nile. It also normalized them as peasants to be developed along western standards. In the process of mapping territories, constructing roads and irrigation canals and integration of local decision making structures into the hierarchy of the state bureaucracy, 'the state', 'private property' and 'law' came to appear as entities with a structure of their own. The introduction of modern science, law and justice did not end the exceptional forms of sovereign rule. On the contrary, as it served to consolidate highly unequal practices of land and water distribution, it created, as Mitchell has it, "a thousand arbitrary powers" (Mitchell 2002, p 77, Boelens 2009).

Today colonial occupation of Nile territory is formally over, but the techno-politics of business networks, governments, development organisation and scientific organisations are very much alive (Verhoeven 2015, Barnes 2014). State funded space agencies work with universities, governments and industry on computerized satellite technologies to calculate crop and water productivities which allow for pixel by pixel comparison of farming entrepreneurs without having to leave the desk of the analyst (Litfin 1997). Not unlike Bentham's panopticon prison which institutionalised a system in which the presence of guards no longer mattered (Foucault 1995 [1977], 199-202), the satellite works to instil a regime of water and crop productivity seemingly independent from those behind the scripts of interpretation. Data democracy, along with open world markets, according to the new liberal ideal, enables individual entrepreneurial subjects to deal with growing climate and market uncertainties.

Yet attention to and awareness of the performative effects of standardization by itself provides no guidance for how to recover responsibilities for remaking the river. As feminist science scholars have been at pains to point out, the qualification of the 'effects' of the power of state, infrastructure, and expertise requires making connections with the ground one seeks to change. Only then can one distinguish 'playful differences' from 'poles of world historical domination' (Haraway 1991, p161, Harding 1986, Smith 1990). Making a difference on the Nile, be it as a water user, engineer or scientist, requires taking sides and making explicit from which position knowledge is constructed  (Star 1990, Haraway 1991). This is not just about the particularities and contingencies that make individuals, locations or frameworks. Positionality is rather about the positioning of oneself and objects/subjects of concern *in the relations* through which these subjects/objects acquire their particular material forms and meanings. This dissertation explores different positions in the remaking of the river, through locating them in the contested relations of work which shaped the Nile morphology since the 19<sup>th</sup> century.

## The morphology of the river as a contested product of work

Over the past decades, various environmental historians and anthropologists have turned to the energies and frictions through which nature is remade to analyse shifting patterns and meanings of global environmental development. Richard White's fascinating account of 'The remaking of the Columbia river', for example, examines how different types of work shaped new forms of cooperation and competition over the Columbia River (White 1995); work by the force of water; and work by different groups of people who manipulated the river for different and at times competing purposes and ways of life. Hugh Raffles' 'In Amazonia' (Raffles 2003) provides a compelling history of how global scientific endeavours produced (and produces) the Amazon as a natural-scientific object. In another book that inspired this dissertation, Anna Tsing mobilizes the concept of 'friction' to understand how imaginaries of development take form in interaction with local particularities of a Kalimantan forest. As "metaphorical image friction reminds us", Tsing suggests, "that heterogeneous and unequal encounters can lead to new arrangements of culture and power" (Tsing 2005, p5). Building on these insights, I view the river as a process of work. My interest is not just in energy, sediments and water flowing, or in the making of friction, but also in the spatial and embodied differences created by the work that makes the river.

This dissertation explores how modern projects of water security solidify uneven positions on the Nile by 'fixing energy in the landscape' (Mitchell 2012, p310). Be it in the form of potential energy of water stored in a reservoir, labour fixed in an irrigation canal, or stereotypes of farmers, women and engineers in agricultural development projects, these fixed forms rearrange the uneven material, economic and cultural possibilities for Nile development in highly particular ways. Marx' analysis of labour as a metabolic process through which 'man' changes nature and 'his own nature' (Marx, 2007 [1867], p198) is an important starting point here. Geographers have built on this insight to understand how the superimposition of capital investments in the landscape produces highly uneven patterns of urban development (Harvey 1982, Smith 1984, Harvey 2001). Major investments in mega-infrastructures such as dams usually take place when capital cannot be productively invested in conventional circuits of capital. These large investments in infrastructure enabled investors to open up new frontiers for appropriation of resources (Merme et al. 2014, Ahlers et al. 2014, Ahlers et al. 2017). Yet as workers in industrial farming would increasingly organize themselves and demand higher wages, and infrastructure would age, the projects that facilitated the appropriation of new resources at one moment would over time become barriers to extraction. Falling profits would produce new crises and make it lucrative for investors again to 'fix' their capital in new infrastructure to expand their horizon of appropriation. Because old structures would not disappear overnight, they would come to stand in the way of new appropriations, making the landscape increasingly sclerotic (Harvey 2001). This created an imperative for capitalists to move the frontier of modernization ever faster, often through reorganising the landscape, drawing new financiers into their remit to organise production at a different scale (Swyngedouw 2007).

By analysing the relations of work through which modern hydraulic development projects rearrange the river, I hope to show these projects are not just a matter of political calculation. By fixing more and more water, people and money along the river for industrial production, the limits to modern water security itself have drastically transformed. The exploitation of cheap labour (e.g. through the employment of wage labour or sharecropping), the appropriation of unpaid labour (e.g. the work of slaves, women, children and sharecroppers which remains unvalued) and the storage of cheap water have not just secured a more reliable production of export agriculture and electricity. These forms of accumulation have also spawned growing inequalities and depleted rivers. These 'external costs' are not just unaccounted for: they are produced and naturalized by the very projects that were built in the name of security (Moore 2015). Rendering water as a 'global commons' to be sustainably managed for present and future generations – as is the popular mantra in water development today (World Bank 2018, FAO 2018, GWP 2015) – often only serves very specific parts of these 'generations': it justifies new reallocations of water at a time when water is increasingly committed by megaprojects that no longer provide returns to the governments and (often Northern) investors that initiated them (Goldman 2005, Merme et al. 2014).

This is not to say that the work of uneven hydraulic development is a singular all compassing political economic project imposed top-down on a powerless river community. On the contrary, projects like mega irrigation schemes and dams currently under construction in Egypt, Sudan and Ethiopia take their specific forms through actions of differently positioned people and infrastructure pursuing interactive and collective projects to engage with the flows of water and sediment. The different positions of these people and objects in changing the river, i.e. their different *agencies*, are shaped by silt particles, property relations and collectives, and meanings that sedimented in the river bed with the work that shaped the river until today. Rather than determining their future, the overlapping patterns of water distribution and values thus embedded in the river provide each water user, engineer or scientist with a range of (conditioned) possible motivations and loyalties for engaging in the rearrangement of the river.

To understand how modern projects of water security change the river and to whose benefit, it is key to analyse how these projects materialize[6] in interaction with physical and cultural forms embedded in the river (cf. Mollinga and Bolding 1996, Van der Zaag 2003, Bolding 2004, Suhardiman 2013). This requires an appreciation of 1. how the modern separation of economic values and moral values is mobilized to justify violent appropriations of land, water and labour (see the section on techno-politics above), and 2. how the outcomes of struggles solidified in the river at one point in time shape specific material positions that come to matter in future rounds of struggle over the river.

With the increasing concentration of people, sediments and historical values attached to reservoirs along the river, the construction of new large diversions of water have become openly contested. This has not only resulted in violent struggles. It also created new collective forms of sharing and developing water. This dissertation turns to a number of modern projects to see how such new patterns are emerging. By doing so I hope to contribute to conversations at IHE-Delft and at development organisations about possibilities of securing waters that are more just and sustainable for more people along the river.

From the above, it should be clear that an inquiry into the intertwining of economics and ethics of modern water security cannot be done in abstract terms in which frameworks of modern water security are formulated (e.g. kilograms food produced, cubic meters water consumed and US dollars gained). To reclaim responsibilities for emerging patterns of water distribution, it is necessary to situate knowledge (Haraway 1997), i.e. to connect knowledge claims to the making of grounded positions in the remaking of the river.

### Making water science and ethics – How a morphology of a river becomes a morphology of water security

A common assumption of modern water security models is that moral accountability is about dealing with *personal* conduct expressed in equal rights, responsibilities and desired behaviour of individual entrepreneurs. As highlighted by Strathern, this has

---

[6] Rather than thinking of the material as the biophysical or the opposite of the ideal, I use Law's definition of *materiality* as "a way of thinking about the material in which it is treated as a continuously enacted relational effect" (Law 2004, p161).

fostered increasing calls for scientists to engage in (self)monitoring, comparison and learning from water users who are adapting to the changing conditions of global development (Strathern 2000). For water management in particular, Pahl-Wostl et al. (2007a) have framed the challenge as "integration of social and content issues ... by relational practices such as task-oriented actions with relational qualities of reciprocity and reflexivity" (Pahl-Wostl et al. 2007a, 4-5). Their focus is on "processes of social learning in which stakeholders at different scales are connected in flexible networks and sufficient social capital and trust is developed to collaborate in a wide range of formal and informal relationships ranging from formal legal structures and contracts to informal voluntary agreements" (Pahl-Wostl et al. 2007a, 11). The answering of this challenge by Nile water management scientists is reflected in a rapid increase in studies documenting, comparing, reflecting and negotiating flows of water through water auditing, citizen science, contingent valuation, agent based modelling, benchmarking and water diplomacy (e.g. Karimi et al. 2012, Tesfaye and Brouwer 2016, Yalew et al. 2016, Abo El-Enein 2011, Kahsay et al. 2015, El-Zain 2007).

Despite this increasing focus on (self)monitoring and reflection on the changing conditions of water management and science, the role of science *in making these conditions* remains largely unquestioned. Remaining faithful to the modern separation of science and morality, many reflexive accounts do not analyse how the objects of their study shape the categories of their analysis. There is a persistent tendency in scientific accounts of Nile resilience to reflect on the impact of case studies of Nile development as if this impact can be innocently read off 'from the field' for analysis, comparison, benchmarking and improvement. Accounting thus remains a matter of monitoring progress against fixed indicators that can be transparently presented to ministers, CEOs and UN commissioners who want directly attributable accounts of the projects whenever questions from parliamentarians, shareholders or member states arise. How to account to those identified by the projects as in need of improvement, development or adaptation often remains a mystery (Turnhout 2018, Hajer and Versteeg 2018, Latour 2018).

To change this situation, a redirection is required from science as a mode of 'discovery' for capacity development of underdeveloped subjects to an interest in 'conversing' with water users to explore and pursue new possibilities for living with Nile water. Here I draw on the feminist scholars of science studies Susan Leigh Star and Donna Haraway, who explore what realigning science and morality entails for objectivity and accountability (Star 1990, Haraway 1991, 1997). According to Star and Haraway, a real connection with (water) users requires taking sides with some of them and not others. In such a science, objectivity is no longer about experiments that can be isolated, transported and replicated anywhere in the world for verification. Instead, objectivity becomes a matter of taking responsibility for positioning instead of denying positioning altogether (Haraway 1991, p196). Only making explicit how knowledge is produced by and for whom, accountability of 'science for development' can be recovered towards those whose waters are redistributed. Both the losses and possibilities that are opened up by this move are considerable: by acknowledging that scientific knowledge (like all knowledges) is partial, it becomes possible to see how particular understandings of water become instrumental to diverting water towards some and away from others.

Whom to converse with, and which links to highlight, become the key questions of any situated knowledge (Haraway 1997, p39). This is not a matter of scientific representation or discursive power. This is about accounting for historical material relations that shape uneven and always changing possibilities of living with Nile waters. To trace the material and historical embedding of these relations, I follow Star's suggestion to start with asking 'who benefits' from modern hydraulic development along the Nile (Star 1990, p43). This requires not just tracing the making of positions of the actors who set the benchmark for success in terms of productivity and efficiency. As Star suggests, the positions of 'those who do not fit' in these categories are at least as interesting (Star 1990, p30). To be unproductive or inefficient in terms of kilograms of crop produced does not imply that people are hapless or outside dominant systems of production. To understand how their at first sight unpromising positions come into being, requires what Haraway calls 'double vision' (Haraway 1991, p154): While from one point of view, one can see how modern projects of water security (re)produce particular relations of domination, from another perspective, infrastructures and water users shape possibilities of river development projects in unexpected ways.

This account of the morphology of the river then, is not about documenting an archaeology of sediments, conducting an ethnography of river life, or a discourse analysis of Nile water security. Instead, this dissertation uses morphology as a methodological heuristic to learn about the uneven historical and material sedimentations of water use practices, knowledges, and soil particles that make the Nile. The connected lessons and experiences from conversations about changing configurations of the soil, water and people yield a partial and tentative account of emerging possibilities for living with Nile water. As a methodological device, as used by Goethe, morphology enables an understanding of how the changing categories used to reorganise the river emerge through the rearranging of the environment. The morphology of the river, in this way becomes a morphology of water security[7].

## 1.4 A morphological account of water security on the Nile

This dissertation analyses how the superimposition of investments in massive dams, irrigation bureaucracies and scientific discourses shape different positions and possibilities of hydraulic development along the Nile. To do so I aim to answer the following research questions:

1.  Who and what make the modern limits to water security on the Nile and why so?
2.  How do projects in the name of water security on the Nile transform the river and who benefits?
3.  What does this imply for the use of science for understanding and changing the emerging patterns of Nile water distribution?

By posing these questions, I aim for a morphological account of securitization of Nile waters that is grounded in the practices that form the Nile river bed, institutions and science. To operationalize such an analysis I build on McMichael's idea of 'incorporated comparison', which explores analytical categories/frameworks - like those of water security - not as given by a "preconceived totality" but through analysis of the "mutual conditioning of its parts" (McMichael 1990: 391). By relating multiple projects over the hydraulic mission era and across the contemporary conjuncture of securitization, I investigate how the categories of modern Nile water security – such as public-private, formal-informal, efficient-inefficient, secure-insecure, just-unjust – are not fixed but emerge with the rearrangement of the river. This is not a deconstruction of techno-politics of scientific discovery. It is about how the shifting rationales and tools of river development are part of changing cultural and material relations through which the river transforms. It appreciates how enormous investments in Nile river development closed off some opportunities for using the river and opened up others (Molle and Wester 2009).

---

[7] This phrase is inspired by Wellmon's - following (Maatsch 2008) - observation that "Goethe's revision of Linnaean taxonomy began as a morphology of plants but became a much broader morphology of knowledge" (Wellmon 2010, p155).

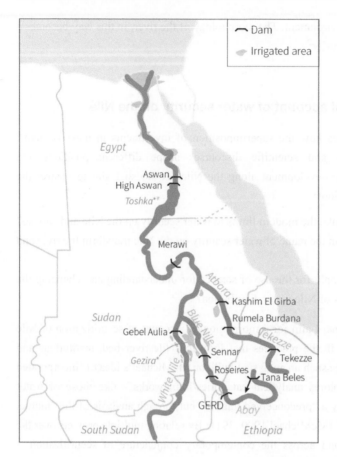

*Figure 1.1: The Nile and its major tributaries and infrastructures that influence its flow north of the Sudd swamps.*[8]

Through focusing on the history of hydraulic development over the Nile's hydraulic mission era and the making of three contemporary hydraulic development projects, I develop a critique of modern water security in four chapters (Chapters 2-5). Each chapter makes an argument about how fluid relations through which actors shape projects of river basin development materialize in techno-institutional networks and discourses of water

---

[8] This maps shows the White Nile north of the Sudd swamps, the Abay/Blue Nile, Tekezze/Atbara and main Nile. The Baro is not displayed and discussed in this paper because no dams with large impact on downstream flows have been constructed here yet. The Equatorial Nile basin, which feeds the White Nile and extends well beyond South Sudan, is not displayed here either. The large evaporation from the Sudd swamps in South Sudan makes that the Equatorial Nile contributes less than 15% of downstream Nile flows and that operation of dams on the Equatorial Nile only have a limited impact on downstream flows until today (Mohamed 2005).

security, and conversely, how these technologies and discourses are used as instruments to forge new patterns of Nile water distribution. The projects are related not only through changing flows of water and sediment in the river, but also through scientific discourses of environmental transformation and relations of food and electricity production that stretch far beyond the Nile.

Chapter 2 historically substantiates this critique by analysing how three waves of massive modern projects of water security shaped the changing river[9]. This chapter draws from (primary and secondary) reports and presentations of stocks and flows of water, sediment and food by colonial and postcolonial engineers and development organisations over the period 1817-2017. The analysis of these documents is necessary to understand how and why subsequent 19th-21st century authors of Nile development mobilized and changed the metrics that justified new projects of modern water security with the changing distribution of water along the river. The focus is on the damming of the river tributaries that carry most of the Nile waters (Figure 1.1). Through tracing how subsequent alliances between local land holders and investors from Europe, the USSR, the US, the Gulf and China rearranged relations by which water, sediment, people and food are distributed along the Nile, the chapter understands the securitization of the river *as a material historical formation* of the Nile morphology and explains the rise of social movements against modern projects of water security.

The chapters thereafter explore how modern Nile water security takes new forms today through analysing the engagements of water users, infrastructure and investors in contemporary projects of water security.

Chapter 3 analyses how the power of the Ethiopian state is inscribed in a hillslope of the Ethiopian highlands through what is arguably the Nile's largest modern hydraulic project in terms of labour: the participatory soil conservation programme in which more than 10 million people are mobilized every year to work on soil conservation for 20 to 40 days. Chapter 4 examines practices of planting of crops, irrigation and canal excavation to recover how a project of Irrigation Management Transfer to water users in the name of efficiency shapes the highly uneven sedimentation of a 95 year old irrigation canal in Sudan. Chapter 5 analyses material and discursive practices of water accounting and irrigation development to understand how knowledges of water accounting now thriving

---

[9]The three waves of large hydraulic development are not particular to the Nile but part of global conjunctures of river basin development since the late 19th century. The three waves are observed and analysed by Warner et al. 2008.

in my own academic department take form with and underline the limited success of a recently constructed high-tech centre-pivot irrigation scheme near Khartoum.

These projects suggested themselves because of the striking visibility of the movements of soil and the similarities in project narratives of environmental security that justified and linked them. When my colleague researcher and I arrived in the village of Yeshat (pseudonym), along one of the tributaries of the Abbay (upper Blue Nile) in 2009, we were struck by the size of two gullies that had grown 8 meter deep, 70 meters wide and more than 150 meters long. When the farmers cultivating the surrounding plots pointed out that the gullies grew as a result of the project of participatory soil conservation, the making of these gullies translated into an important research focus.

A year later, and 700 km down the river, we learned not only that the Gezira 'plain' had been burdened with enormous piles of sediment along irrigation canals. People in the century old Gezira irrigation scheme highlighted that recent efforts by the government to transfer canal maintenance to tenants had turned sediments into a lucrative business at the detriment of reliability of water supply.

The opposite bank of the Blue Nile showed signs of fresh movements of large volumes of soil too: a similar grid of irrigation canals and silt piles replaced by a new centre-pivot irrigation scheme where alfalfa is grown for export to the Gulf. While this new scheme is hailed by Sudanese and foreign officials and development organisations as 'the model' for future Nile agriculture, efforts to 'scale up' this model did not materialize as expected. The boundaries of the new farm did not expand but were fortified and Waha (oasis - Arabic) remained a seemingly isolated promise of modern Nile development.

These experiences triggered our interest to analyse how these projects related through flows of sediment and water and networks of science and investment. A focus on narratives of water development, relations of agricultural investment and the form of the river enables an analysis of how these projects are conditioned by and shape different geographical positions and "distinct and overlapping 'social times'" (McMichael 1990, p360) along the Nile. By analysing the projects as related pieces of a larger whole, this dissertation aims to understand how distributions of Nile water, sediments and science develop and materialise together.

Between 2010 and 2017, I spent 21 months in the Choke Mountains of Ethiopia, and in the Gezira and Waha irrigation schemes in Sudan with researchers and students from Ethiopia, Sudan, the Netherlands and the UK. Chapters 3-5 draw from a) observations of human activities and the functioning of irrigation and drainage technologies on a 50 ha hillslope in the Choke Mountains (Chapter 3), 800 ha of land supplied by one of Gezira's minor canals (Chapter 4) and 7 'networks' of investors in cultivation around the Waha irrigation scheme (Chapter 5); b) numerous conversations and 250 semi-structured interviews with people involved in cultivation of the land, operation of water infrastructures, trade of crops, and implementation of development projects; c) Crop maps, market surveys, irrigation/drainage maps; d) an analysis  of changes in the landscape based on discussions of aerial photographs and a satellite images of the two gullies we studied (Chapter 3) and the irrigation schemes at Waha (Chapter 5); and e) participation in meetings and trainings of religious networks, government programmes, and other social networks. Apart from that, these chapters draw from discussions, meetings in offices and lecture halls of development organisations, governments, private enterprises and universities involved in Nile development between 2008 and 2017.

Both the research and my understanding of the projects is profoundly shaped by a) social and material relations carved into the Nile landscape by earlier projects of hydraulic development[10], b) the high government stakes in the 'renaissance' of Nile development claimed in both Sudan and Ethiopia[11], and c) expectations of the donors and government bodies that shaped the research and of the people in the project areas[12].

From the outset it was clear that the projects of our interest were politically sensitive, and they still are. Our obvious identification as outsiders and the connections with officials to gain access to the project areas, shaped our interactions in the project areas in influential ways. The violent history of distributions of land, conscription of labour and forced migration by projects of development and the ongoing repercussions for being labelled as 'against development' made many in the project areas reluctant to discuss the projects of our interest.  Moreover, the absence of questionnaires and sophisticated measurement equipment created deep suspicions about our 'expertise as researchers' amongst the security agents we encountered. This created considerable risks for the researchers working with me: if these suspicions were documented in security reports, this might have severe career implications for them.

---

[10] For valuable contributions that bring into view part of this history in Ethiopia: see James et al 2002, Dessalegn Rahmato 2009, Pankhurst and Piguet 2009. For valuable contributions that bring into view part of this history in Sudan, see Ali 1989, El-Battahani 1999 and Abd Elkreem2018

[11] For valuable contributions on the hydraulic renaissance in Ethiopia and Sudan see Cascão and Nicol (2016) and Verhoeven (2015)

[12] For a thoughtful contribution on the history of the links between donor community and the Ethiopian bureaucracy see Fantini and Puddu (2016)

This constrained research context is at the root of several considerations about the setup of this research. Questionnaires and consent forms that are now gaining ground in social science research do not only reproduce the modern scientific separation between the researcher and subjects handing over information (Haraway 1991, p,197-198, Strathern 2000). In the context we worked in, we suspected these may also yield short and socially desired answers that would have reduced the potential richness and complexity inherent to the relations of water flows, investment and science. Instead of seeking for 'signed off' consent – and with agreement of, but considerable risk to, my co-researchers – therefore a lot of time was invested in gaining trust and explaining the questions driving the projects and the reasons for these in universities and development organisations. This did not prevent a good deal of scepticism about the endeavour. To protect everyone who informed the analysis presented here from undesirable consequences of publication, I have used – as promised – pseudonyms, also for some of the locations. Moreover, I take full responsibility for the analysis in this dissertation.

This research not only created a moral duty to work hard to return something for what so many in Ethiopia and Sudan gave to make this dissertation possible. The violent histories and remaining scepticism conveyed to me by my 'teachers of development' provided constant reminders of the polluted roots and the limitations of the colonial academic apparatus of which this dissertation is part. In Chapters 3-5 I hope to at least make clear that the projects of Nile development analysed in this dissertation are not intended as representative of some imagined 'Nile population in need of development'. On the contrary, building on the historical understanding of the securitization of Nile River in chapter 2, chapters 3-5 bring to the fore the agencies of water users and infrastructures which make that water is never securitized as imagined by colonial engineers of Nile development.

Chapter 6 concludes the dissertation by drawing together how modern indicators and principles of Nile water security are structured by, and (in)formed, the changing morphology of the Nile river, its institutions and knowledge. Bringing into conversation a. how the superimposition of colonial grid justifies violent appropriations of land, water and labour and b. the varied agencies of water users and infrastructures in shaping new forms of water distribution and cultivation, the dissertation shows how the prospects for new megaprojects are reducing now the environmental and political costs of diverting water to new hypermodern schemes are rising. The targeted subjects of modern river development make use of the new spaces for 'river development' thus created by carving out their own projects. They creatively mobilize old irrigation and drainage infrastructures in ways that escape the universal logic of modern water security.

# 2

# Materializing water security – two hundred years of modern hydraulic development on the Nile (1817-2017)

## 2.1    Introduction

In 2011 operators at Sennar dam diverted a record 9 billion $m^3$ water (MoIWR 2012) - around a $10^{th}$ of the entire Nile flow - to the century old Gezira irrigation scheme in Sudan (Figure 1.1, p20). Both within the country and outside, this diversion is observed with growing suspicion. Politicians, policy makers, investors and engineers ask if it is not possible to find a better use for the water than sorghum cultivation through furrow irrigation (Elnour and Elamin 2014, Awulachew et al 2010). Their questions are growing louder as billions of dollars are invested in new high tech irrigation schemes and hydroelectric dams. Pivot irrigation schemes in Sudan and Egypt have started growing alfalfa for export to Saudi Arabia. Hydro-electric dams in Sudan have quadrupled the country's electricity production over the past decade. In Ethiopia four new sugar factories and a gigantic hydropower dam are under construction along the Blue Nile and tributaries. With rivers better regulated by reservoirs and more efficient irrigation systems – the popular logic goes – the sum of food and energy produced with Nile waters can be greatly increased (NBI 2009, p9). With the flow of the Nile ever more concentrated in and regulated by reservoirs, the production of food and energy along the Nile is indeed increasing (NBI 2012). Yet this has not reduced concerns of many along the Nile who struggle to access water, energy and food. On the contrary; as the new projects resonate with earlier struggles over territory, ethnic conflicts and international debt, conflicts over water along the Nile are on the rise.

Growing investments in large scale production of food and green energy are at the core of the work of international development organisations. Working with governments around the world, they share a broad concern to attain food and water security through intensification of production, valuation of ecosystems services and putting in place good structures for their management. Yet despite the focus on intensification of production, valuation *and distribution* of the benefits thereof, the concepts and indicators that are subsequently mobilized give little guidance for analysing how intensification of agriculture and hydropower production materializes and for whom. Water is rendered an abstract(able) resource that can be measured, converted, valued and interchanged in terms of $m^3$, kg, US$ and coefficients thereof; equally abstract communities are to be prepared for unknown risks based on simulated scenarios of growing populations and increasing droughts. This prepares the ground for numerous policy reports and scientific articles and Sustainable Development Indicators that propose new technical measures to liberate countries remaining "hostage to hydrology" (Grey and Sadoff 2007, p545) and adaptive water governance to better cope with climate uncertainties (Pahl-Wostl et al. 2012). But what if the limits to water security are not only the reason for but also the product of successive projects of modern hydraulic construction and governance?

Rather than searching for more indicators, thresholds or frameworks that help prepare for future scenarios of water crisis, this chapter argues that no limits of water security are independent from the historical organisation of the river. To ground our analysis of possibilities for reorganising the Nile river through contemporary water development projects in the following chapters, this chapter explores how large projects of river basin development that were implemented in the name of security changed the Nile and the ways it connected and divided people along the river and across the world. This takes us back to the French engineers who were hired by Mohammed Ali in 1817 to design barrages over the Nile to raise the water above the delta and thus enable the perennial irrigation of cotton. Their *modern* sense of water security rendered the river as a natural resource to be rationally developed by and for Mankind. This 'hydraulic mission' became the rationale of the vast amount of *modern* scientific studies that aim to contribute to developing the Nile. From the late 19th century the representation of the river as a 'natural flow' – commonly depicted in a hydrograph with an extreme annual variation (Figure 2.1) - would become central to the calculation and legitimization of Nile development projects.

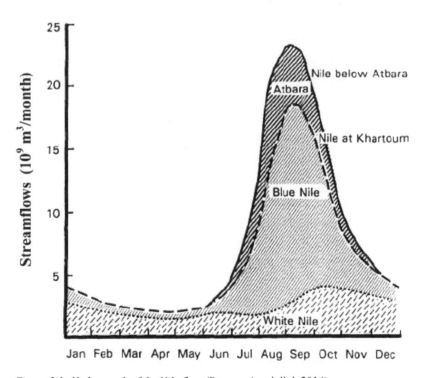

*Figure 2.1: Hydrograph of the Nile flow (Source: Awadallah 2014)*

As highlighted by the most acclaimed British Nile engineer of the late 19th century, William Willcocks:

> "In Mohamed Ali's time, [1805 - 1848] the great preoccupation of the [Ottoman-Egyptian] Government was the pressing on of the cultivation of cotton, and as this crop needed perennial irrigation, the securing of an abundant supply of water all the year round was the problem of the day."
> (Willcocks 1904, p73)

This did not mean that no Nile water was captured or manipulated until then. On the contrary, the elision of earlier manipulations of nature was exactly the point of modern hydraulic development. As highlighted by accounts of land and water use practices in Egypt by Mitchell and Barnes, the rendering of the river as a natural object to be harnessed through human ingenuity was the result of continuous political/material work to reform the existing river bed and the institutions of its management (Mitchell 2002, Barnes 2014). To understand *how* and *for whom* particular projects of modern securitization of the Nile were a success, this chapter aims to extend their analysis by providing a history of how modern water security *materialized* through large scale hydraulic engineering. My specific interest here is in how successive waves of Nile development projects in the name of water security created a particular trajectory of Nile development. Three waves of massive investment in modern hydraulic development over the past two centuries each sought to organize water at a particular scale and thus produced the very uneven distributions of water and labour that came to undermine the imperial, national and global models for water security in which name they were instilled. [13] As we can witness on the Nile today, the old dams and bureaucracies have not been simply removed. Tracing carefully what happens to old infrastructure as the modernization of the river progresses, I examine how subsequent modern development projects have changed the river, and how water security has shifted along the way.

In the following sections I analyse how the three waves of hydraulic construction rearranged relations between investors, states and associations of water users along the Nile. Bringing into conversation the modern calculations of water security that were mobilized to arrange control over land, water and people with accounts of specific changes in the forms and meanings of the Nile, the chapter explores how each wave produced its own Nile calculations and ecohistorical limits. By tracing how each wave of

---

[13] The three waves of large hydraulic development are not particular to the Nile but part of global conjunctures of river basin development since the late 19th century. The three waves are observed and analysed by Warner et al. 2008.

development thus posed material conditions for subsequent hydraulic projects, I aim to get a grip on the possibilities for rearranging the Nile today.

## 2.2    Colonizing the Nile Basin - river basin control for cotton cultivation 1817-1937

After millennia of accretion of mud that originated from the Ethiopian highlands, the growth of the Nile Delta in Egypt was suddenly reversed in the 19th century (Mikhail 2011). This turnaround was all about perennial irrigation for cotton cultivation. Before this shift, crops were cultivated on soaked lands on which the receding annual flood had left fresh sediments. The long growing season of cotton required tapping into the summer water[14] before the flood arrived. To enable this shift the Ottoman rulers conscripted labourers to construct barrages and embankments on and along the Nile to raise the 'summer water' above the landscape for irrigation and protect the crops against the 'flood' water that would follow. Whereas before, the flood would turn the delta into a vast lake, the barrages, banks and canals that enabled perennial irrigation 'inverted' the hydrology of the delta and kept it dry when floods were arriving (Hamdan 1972, p127, Brown 1896). In less than a century cultivation in almost the entire delta was transformed from flood irrigation to perennial irrigation. With the increasing storage and consumption of water, the supply of sediments to the delta was severely reduced and the shoreline started retreating.

Perennial irrigation was far from just an Egyptian affair. Grains levied from peasants were sold on European markets to finance the manufactured goods and labour for construction of new canals, pumps and railways. The conscription of Egyptian farmers, the Ottoman conquest of Sudan for providing slaves (1821), and the forced production of cotton would tie millions Egyptians and Sudanese into the British imperial economy. Not only would cotton cultivation make farmers themselves increasingly dependent on food and fertilizer from foreign markets; to finance the expansion of cotton infrastructure, the Ottoman rulers sold land to European investors and turned communal lands into private property that could be taxed (Tignor 1966, Mitchell 2002). Through heavy taxation, the expansion of Egyptian irrigated agriculture from 840 thousand hectare to 2 million hectare over the 19th century was mortgaged on an increasing peasant indebtedness (Richards 1982, p39). Failing to pay the debt, more and more smallholders saw their lands appropriated by

---

[14] The summer season is the season before the Nile flood arrives in July. It stretches from March to July, during which the cotton is planted.

national and foreign estate holders, who let them produce subsistence crops on part of the land in exchange of their labour on cotton fields. When a global economic crisis hit in the 1870s and mass revolt threatened European investments, the British army took over and occupied Egypt in 1882. The decades that followed would bring new dams and irrigation schemes to the Nile at an unprecedented scale[15].

British expansion of cotton production and control over Egypt and Sudan were justified by new western hydraulic calculations. For centuries water level gauges along the Nile had been used to organize cultivation based on the height of the annual flood. From that moment, the statistics of earlier recordings were used to reorganise the river. Planners and engineers imagined to tame the river through increasing summer supply and reducing floods. Distribution of water was no longer justified by the sovereign's approval or local arrangements but conceived of in terms of land productivity and efficiency. Mapping of land and plotting of hydrographs and water balances was crucial not only to control the river and its people. The large-scale transformation to perennial irrigation not only enabled the production of cotton but also institutionalized a hydro-agricultural bureaucracy that gained increasing control over the Egyptian peasantry, and that was justified in terms of social equity. As a British engineer described it, the "system of [perennial] irrigation by rotation is of great advantage, not only in checking the loss of water in the channels, but in teaching economical irrigation of the cultivators and in ensuring an equitable division of supply among the people" (R. B. Buckley 1893, p237 quoted by Cookson-Hills 2013, p205).

The massive expansion of perennial irrigation for cotton production created modern concerns of water shortage. No longer was cultivation dependent on the height of the Nile flood, instead it became dependent on volume of summer water available for cotton cultivation. As the cotton season stretched from March to October, the area that could be irrigated was limited by the amount of water in the Nile before the flood season started in July. Because storage of large amounts of water on the flat Delta plain would require large reservoirs which would be costly to construct, take up valuable space and evaporate a lot of water, the basin was declared as the appropriate and 'natural' scale for organizing the river. As early as 1894, Willcocks - who would later be in charge of designing the Aswan dam - reported that upstream dams were required because storage in Egypt would be insufficient (Willcocks 1894). A British journalist in Egypt framed it as follows in 1904:

---

[15] For a fascinating account of the Nile politics in the age of the British occupation of Nile basin territory see Tvedt 2004.

"all Nileland is one country. No divided sovereignty is possible; there must be one firm hand over all... Looking back, it is easy to see how it all followed in logical sequence. Everything depended on the Nile. The more Egypt was developed, the greater grew the need for the regulation of the Water. The rulers of Egypt need have troubled little about the fate of countries divided from them by so many leagues of rainless desert, but for the link of the all-important river" (Peel 1904 p134-135).

Perennial irrigation for cotton cultivation thus not only required the inversion of the Nile hydrology, it produced a new geography of water security that rested on control over the entire basin to ensure a continued and augmented summer supply to Egypt. To prevent occupation of the upper Nile territory by other European colonizers, London ordered the conquest of Madhist Sudan in 1898. Soon after the conquest of Sudan, British engineers set out for new explorations of the upper basin. The resulting 'Report upon the basin of the Nile with proposals for the improvement of that river' (Garstin and Dupuis 1904) identified projects for regulating the upper river to augment the supply of summer water to Egypt. The report included a barrage at the outlet of Lake Tana in Ethiopia, dams on the Blue Nile, and the construction of a new 340 km canal from Bor to the Sobat[16] which would reduce the evaporation of water from the Sudd swamps in the Equatorial Nile region. Several storage dams in Upper Egypt and Sudan would capture water at the end of the flood season for use during the remainder of the cotton growing season and for planting of the new crop before the next season. They would be emptied at the beginning of the flood season to prevent that high sediment loads during the beginning of the flood season would deposit and fill the reservoirs with silt. During the decades that followed, British officials in the Egyptian irrigation ministry would expand and supervise the partial implementation of the proposals to control the Nile *basin* (Willcocks 1919, MacDonald 1920, Newhouse 1939, Hurst, Black and Simaika 1946, Hurst 1952) (Figure 2.2).

This was not just mere scalar imagination or paper engineering logic. The idea of the basin being the optimal scale for what Willcocks called "securing an abundant supply of water all the year round" (Willcocks 1904, p73) took hold in the conquest of the Sudan and the construction of annual storage dams. Convinced by the profits and calculations of the modern economy and hydrology, investors who saw European markets faltering again in the first decade of the 20th century diverted increasing amounts of capital to the

---

[16] Later named the Jonglei canal - see fig 2.

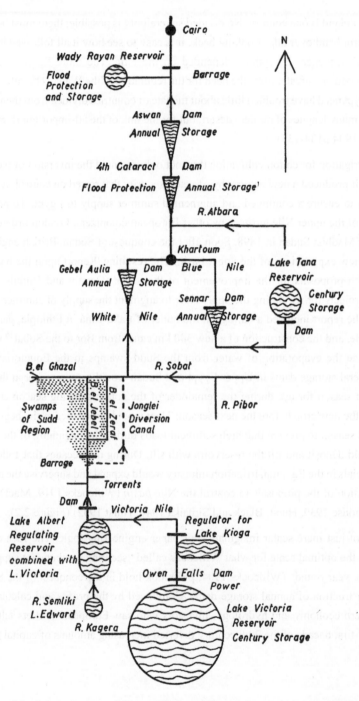

*Figure 2.2: Schematization of Hurst, Black and Simaika's century storage scheme for water supply to Egypt (1946). (Source Hurst 1957, p282)*

Nile. The 1902 Aswan dam, the first dam on the main stem of the Nile, increased the average 'summer supply' by 50% and thereby enabled an increase of perennial irrigation by 210,000 hectares (Willcocks 1904, p75-76). Through the reservoir, Egypt's position as Britain's main supplier of long staple cotton was secured. The dam provided engineers and investors with the 'objective' evidence for the success of modern hydraulic development. The British journalist Peel wrote:

> "Well might Lord Cromer and the irrigation engineers review their work with satisfaction. To them, the [Aswan] Reservoir means the successful culmination of a great policy long and steadily pursued, nothing less than the establishment of the prosperity of Egypt upon a sure and certain basis: for this is what the regulation of the Nile involves. In Egypt, at any rate, they require no formal monument. Their praise stands clearly writ on the face of every cultivated field throughout the country" (Peel 1904, p100).

Indeed the dam significantly contributed to rising cotton production levels at the start of the 20th century and mitigated the drought of 1905. Yet the progress of perennial irrigation never materialized for most involved in cotton cultivation. On the contrary, many smallholders found themselves in periodic debt to finance fertilizers and additional labour for the weeding and picking of cotton. To repay their loans quickly, many of them reduced fallowing by shifting from a three-year to a two-year rotation (Richards 1982, p80-83). This contributed to the increase in use of irrigation water and, with the absence of floods to flush the land, led to an accumulation of salts carried with the water in the soil: the major cause identified by British engineers for yields to drop steadily from over 100 kg/ha in 1895-1899 to less than 80 kg/ha in 1905-1909 (Willcocks and Craig 1913, p418). While total cotton production in Egypt doubled between 1885 and 1910, by 1919 more than a third of the peasants was landless and more than 70% of Egyptian cotton was produced on estates (Tignor 1966, 226-234).

As the increasing landless population came to threaten the imperial force that had intervened in Egypt, industrial cotton cultivation would not only organize the limits of control of summer water through conquest of the river basin, but also spark the rearrangement of these limits. Facing rising opposition and a global economic recession, the British government declared Egypt's independence in 1922. To protect its interests in the region (notably in capital invested in cotton production and the Suez Canal), it decided to retain control over Sudan and with it the water supply to Egypt. The expansion of the gigantic Gezira scheme in Sudan - originally planned to draw only non-summer water - was now seen in Egypt as a direct threat to Egyptian water supply. Following growing outrage over irrigation development in Sudan, the Egyptian and Sudanese governments installed a Nile Commission to relief tensions in 1925. In an exchange of notes which came to be known as the 1929 Nile waters Agreement, the UK and Egypt agreed that Sudan was allowed to abstract water from the Sennar reservoir, but only if it would not compromise Egypt's summer water supply. While basin control of summer water for Egypt was thus secured on paper and embedded in the annual storage dams built on the Nile at Aswan, Sennar, and later Gebel Aulia, the material reality of dam development of the 1930s marked the beginning of a shift towards a different model for organizing the river. No longer wanting to rely on Sudan for storage of Nile water, the Egyptian parliament decided to heighten the [original] Aswan dam and double its capacity before agreeing to the next upstream reservoir at Gebel Aulia in Sudan (see Figure 2.3). Holistic river basin planning had thus been dented. The next wave of dam construction on the Nile would take the form of *national* development.

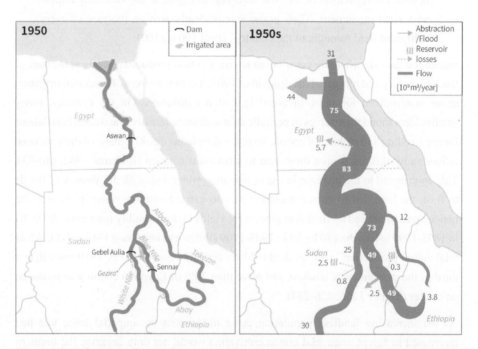

*Figure 2.3: Left: Dams after the first wave of dam construction on the Nile (1899-1937) Right: an estimation of distribution of the flow, abstractions and reservoir evaporation after the first wave of large scale hydraulic construction in an 'average' hydrological year. Note: Gebel Aulia dam was owned by Egypt until 10 years after the Aswan High Dam was constructed (until 1977) and increased the summer storage for Egypt which allowed Sudan to withdraw more water from Sennar Dam to the Gezira. While the Aswan High Dam has made this original function of Gebel Aulia dam obsolete, around 2.5 BCM/a evaporates from the reservoir until today. Sources for flow diagrams in this chapter: Sutcliffe and Parks (1999), Chesworth (1994), Whittington and Guariso (1983), MoI, UNDP, IBRD (1980), Blackmore and Whittington (2009), SMEC (2012), ENTRO (2015) and Molle et al. (2018).*

## 2.3. The Blue Revolution (1952 – 1973) – Expanding the agro-industrial complex for 'national' development

Whereas the first wave of Nile dams was driven by territorial conquest in name of cotton production, the second wave would capture multiple years of Nile water in large 'national' reservoirs to reclaim 'new lands' from the desert in Egypt and Sudan. As we will see, the large sums of money and particular expertise required for these megaprojects of ´national development´ would provide the basis for the creation of new international dependencies.

Since the 1930s, Egyptian nationalists had promoted social welfare plans to develop new model villages that would provide the basis for expanding agriculture into the desert (El Shakry 2007). By the early 1950s these had done little to ameliorate the situation of millions of landless people . When rising food prices in the aftermath of the Second World War led to famine conditions, a group of army officers calling themselves the Free Officers Movement took control over Egypt in 1952. One of the first orders they gave was to re-distribute large estate lands and seek for new economic alliances to finance projects for its national development. But the Free Officers were careful not to alienate themselves from the Egyptian middle class of small estate holders. Putting the threshold for redistribution high (for Egyptian standards) at 50 acres, they left 80% of the cultivated land untouched and imagined real reform to take place through large-scale irrigation development in the desert. The Tahrir project with a planned area of half a million hectares was presented as a new beginning for modernizing Egypt in 1955. Disappointing results - three years later only 5,500 hectares were cultivated by 389 families (Voll 1980) - did not make the government change its model for development. On the contrary, as control over water in the Upper Nile *territory* was no longer guaranteed, it highlighted that the construction of a gigantic Aswan High Dam was required to capture all *water* within Egypt's national boundaries and expand cultivation of the desert.

In Sudan, still under control of the British-Egyptian condominium government, pressures to secure water for expanding irrigation were also rising. Seeking for continued loyalty of rich cotton tenants after independence (1956), the government started to issue pump licences to religious and political leaders and businessmen. Many of them were Gezira tenants and money lenders who got rich when cotton prices had jumped during the Korean war of the early 1950s and sought to reinvest their money beyond the colonial canal grid. Through issuing 1,557 pump licenses for irrigating an additional area of 260,000 hectare under the Nile Pump Control Ordinance, almost half of Sudan's irrigated area at the time, was put under private pump irrigation (IBRD 1959, Niblock 1987, p32).[17] When Sudan became independent and formed its first government in 1956, 65 out of 76 members of the National Assembly owned or invested in private cotton schemes along the Nile (Ali 1989, p119). While the Sennar and Gebel Aulia dams constructed on the Blue and White Nile stored enough water to cultivate the Gezira scheme without touching Egypt's Summer water as stipulated in the 1929 agreement, the new private pump schemes would soon put Sudan's storage of summer water under pressure.

Not only did the expansion of irrigation shift the attention from unimpeded passage of 'summer water' of the White Nile to over year storage of the 'flood waters' of the Blue Nile, the search for finance for new dams would draw Sudan and Egypt squarely into cold war politics. Within a few years after the 1952 revolution Egypt's General Nasser discussed the financing of a gigantic dam to store Nile waters in its own territory with the World Bank, the UK and the US, which all vied for extending their political influence in the region. Initially Nasser and Western partners seemed to be heading for a deal. When Nasser increased Egyptian ties with Russia and bought arms from the Czech Republic, the US and the UK, withdrew the offer. This was more than a cold war signal. It proved a crucial move in undermining UKs waning colonial interests for business in the Nile and Africa (Tvedt 2004). In response to the pull out, Nasser nationalised the Suez Canal and claimed Egypt would build the dam without Western support. A few years later, and much to the surprise of the Western world, Nasser managed to secure a 400 million Rouble Russian loan for the first phase of dam construction (Fahim 1981, p58). What remained to be solved was a new agreement over sharing Nile waters.

The revolutionary government of Egypt had been pushing Sudan for a new Nile agreement ever since it came to power in 1952. The reservoir behind the proposed High Dam would flood more than 150 km into Sudanese territory and submerge Sudan's main northern town of Wadi Haifa. More importantly, Sudanese party leadership had its own aspirations for Nile development. Its investment in pump irrigation turned it diametrically against a new water deal with Egypt. When Sudan refused to sign a new deal, Egypt withdrew the support of its national bank, which still provided the currency for Sudan. The Sudanese government responded through opening the first stage of the 420,000 hectare Managil extension to the Gezira scheme, arguing that the 1929 agreement was no longer valid as it had been signed by the United Kingdom on behalf of Sudan (Ali 1989). The Egyptian boycott and the further deterioration of the economy led to uprisings

---

[17] The pump schemes employed a large number of cultivators. Ali reports some 400,000 of them on the pump schemes along the White Nile around Kosti alone (Ali 1989, p93).

organised by the Unions of Sudan. To prevent a coup, the Sudanese Prime Minister Abdallah Khalil invited the Sudanese army to take over the country in 1958.

Within a month, the Sudanese military government accepted the US 'Food for Peace' programme. The USA were keen to provide aid to Sudan in exchange for a foot on the ground in the region. They sold cheap American wheat on the Sudanese market and promised support to the Managil extension if a deal with Egypt on the Nile waters would be reached (Ali 1989, p149). Within a year two generals signed the Agreement between the Republic of the Sudan and the United Arab Republic for the 'Full Utilisation of the Nile Waters' (PJTC 1959). The deal detailed the division of current 'unutilized' water through projects as follows:

1.  UAR would construct the Aswan High Dam for over-year storage in Egypt.
2.  Sudan would construct the Roseires dam and any other works to utilise its share of Nile waters.
3.  The definition of the current Nile flow at Aswan at 84 BCM/year, from which 10 BCM/year would be deducted as this would be lost from the Aswan High Dam Reservoir through seepage and evaporation.
4.  The division of the 22 BCM 'net benefit' from the Aswan High dam[18] "of which the share of the Republic of the Sudan shall be 14.5 Milliards and the share of the United Arab Republic shall be 7.5 Milliards" (p3). Together with the "acquired rights" (p3, which were determined through converting the 1929 agreement into a flow of 48 BCM/year for UAR and 4 BCM/year for Sudan), the new allocation was set at 18.5 BCM/year for Sudan and 55.5 BCM/year for Egypt.

While gigantic reservoirs at the upstream borders of Egypt and Sudan marked the shift from basin storage to storage for national development of these countries, upstream countries were not consulted about the dams or the agreement. When Ethiopia's imperial government saw itself excluded from another Nile deal, it allied with the US Eisenhower administration, which "saw conflicts over the development of the entire Nile River basin as a key component of their strategy to exercise influence over Egypt's president Gamal Abdel Nasser" (Sneddon 2015, p89). It was thus no coincidence that while Egypt and Sudan negotiated the 1959 *agreement for full utilization of the Nile waters*, the US Bureau of Reclamation engaged with the Ethiopian government in a study of development of the upper part of the Blue Nile. The study identified four locations – one of which is discussed

---

[18] i.e. the assumed total flow of 84 BCM/year minus the assumed 10 BCM/year losses minus the 52 BCM/year commitments as agreed in the 1929 agreement

in the next section - for the construction of hydropower dams and the development of 430,000 ha of irrigation (USBR 1964). Yet by the time the study was completed in 1964 the US was careful not to finance implementation, as Egypt had evolved into a pivot ally of the US in the Middle East. After the Aswan High Dam debacle, the US had regained influence in Egypt through the provision of cheap wheat to feed Egypt's growing landless population.[19] With the Ethiopian empire lower on the geopolitical agenda and struck by civil war for two decades from the 1970s, plans for dams on the upper Blue Nile remained in government desks for another three decades.

While the hydraulic mission in Ethiopia was thus stalled, the 1959 agreement paved the way for renewing Western influence downstream along the Nile. In Sudan, the World Bank, West Germany and Italy provided support for a new round of investments in Sudanese public irrigation infrastructure. With the 1959 agreement in place, 70% of the Overseas Development Assistance under Sudan's Ten Year Plan of 1961-1970 was invested in the Managil extension of the Gezira scheme, Roseires dam, Kashm el Girma dam and two new irrigated sugar schemes (IBRD 1963). The dams would not only enable the expansion of irrigation - the increase in water stored in the Roseires reservoir and the import of machinery and fertilizers from the West enabled the intensification of existing irrigation. In the decade after the opening of Roseires Dam, water consumption in Sudan doubled from 3 BCM to 6 BCM per year. Meanwhile cotton prices had fallen and Sudanese owners of pump schemes shifted their money to mechanised farming of sorghum on large tracts of rain fed land. This was facilitated by the Sudanese government, which bought out all medium and large size pump irrigation schemes at favourable terms in 1968 (Hewett 1989, p211).

While the Gezira tenants were not bailed out by the government, the fact that the scheme still produced more than half of the country's foreign currency earnings gave them considerable leverage. In 1946 a strike had for the first time demonstrated the power of their organisation. More than 75% of the tenants refused to plant the cotton crop for six

---

[19] Much to the dislike of the Ethiopian Emperor, American support to Ethiopia's water development remained limited to a small dam for a sugar factory on a tributary of the Nile at Fincha'a. Bitterly he remarked about the USBR's involvement in hydraulic development that "It took them five years to make a survey… Meanwhile you have doubtless heard of the big fuss further down the Nile where the Russians dedicated a much bigger dam which they built not in 40 years but in about five years" (Emperor Haile Selassie quoted by Sneddon 2015, p98). As the Ethiopian empire was increasingly prone to internal unrest and overthrown by a so-called socialist Dergue government in 1974, which turned to civil war for almost two decades, plans for dams remained unimplemented.

weeks and successfully negotiated a larger share of the cotton profit (El Amin 2002, p164). Over the decades that followed, their power would grow. By the 1970s the gigantic colonial irrigation grid of the Gezira had thus not only institutionalized a highly divided irrigation society but also produced one of Africa's most powerful unions. In 1969 the tenant Union (along with the railroad/workers union) brought (later) president Jafaar Nimeri to power. He nationalized the Gezira in 1970 and his domestic food policy initially alienated him from local elites. The increasing bill to repay the loans for hydraulic infrastructure and war in the south eroded his position. After the union turned against him, he eliminated its leaders and re-aligned his interests with national elites and investors from the Gulf and the West (Niblock 1987).

Capitalising on the desire of the Gulf States to become independent from food from the west after the Arab–Israeli wars, Nimeri worked with them to turn Sudan into the 'bread basket of the Arab world' (El-Battahani 1999, Woertz 2013). Over the five years that followed, Kuwait and Saudi Arabia invested billions of dollars of oil money in rainfed cultivation of wheat and sugar production for commercial export. The World Bank facilitated the redirection of investments from (post)colonial irrigation schemes with hundreds of thousands of smallholder to large rainfed mechanized farms (IBRD 1968, 1972)[20]. Two decades later, and after the 'conversion' of almost 4 million hectares to mechanized farming by 8,000 (mostly absentee) landowners,[21] it became clear that this had produced a human and environmental disaster. Most investors mined the soil in a couple of years and then moved on, and never produced enough to generate significant earnings from export. The rising number of displaced people came to rely on sorghum markets that were increasingly controlled by Saudi supported Islamic banks that were set up for implementation of the bread basket strategy (El-Battahani and Woodward 2013)[22]. Only 14 years after the construction of the Roseires Dam in the name of food security and export earnings, the country was bankrupt. Wheat imports doubled to around 500 million

---

[20] The nationalized irrigation schemes like the Gezira were increasingly pushed to substitute for the imports. With cotton acreage reduced and the maintenance paid from cotton proceeds declining, and silt intake increasing by the early intake of water to accommodate the food crops Gezira faltered. By the end of the 1970s the Gezira was portrayed by international organisations as a system with a serious maintenance crisis and an inefficient bureaucracy (World Bank 1980).
[21] Roughly the same area of land as cultivated by the 2.5 million families in Sudan which relied on rainfed farming at the time (Duffield 1990, p191).
[22] By the early 1980s they controlled around 40% of the sorghum stocks (Woertz 2013, p189). While Sudan's most influential bank, the Faisal Islamic Bank paid 90% dividend on investments over the 1979-1982 period (Shaaeldin and Brown 1988, p135), Sudan's Northern region faced increasing food insecurity from 1982.

tonnes per year (FAOSTAT 2017). When the government turned to the IMF to reschedule its loans, the latter insisted on the removal of subsidies on fuel and bread after which oil and bread prices increased by 66% and 75% respectively (Brown 1988). Many who lost their land moved to Khartoum when they found no food in rural markets in 1984. While this contributed to the popular revolt under the leadership of the Unions which removed president Numeri in 1985, an estimated 240,000 died in Sudan from famine (De Waal 2005).

At that time there was only one country which received more US AID than Sudan: Egypt. Not only did the US reward Egypt with development support in exchange for peace with Israel, by the 1980s Egypt had developed into one of the world's largest consumers of US wheat. This was not the result of falling agricultural output or a growing population. On the contrary, the gigantic volume of water stored at Aswan protected the country from the droughts and enabled a rise of grain production that was larger than the rise in population in the 1970s and 1980s. With western support 300,000 hectares in Upper Egypt had been converted to perennial irrigation after the construction of the Aswan High Dam. Intensification of cultivation in the delta enabled the increase of the cropped area by another 400,000 hectares (MoA Egypt et al 1983). Yet this was of little benefit to smallholder cultivators, who were forced to produce cotton, sugar and rice and sell it to the state below world market prices (World Bank 1979, p40). While smallholder farmers struggled to produce their 'quota' and became dependent on food aid from the US, large landholders had no difficulties to comply with the quota and used the rest of their land to produce fodder to tailor to the growing demand for meat by the Egypt upper class. While in 1985 almost half of the grains produced in Egypt were fed to animals, the Egyptian population developed into a rapidly growing market for US grain producers (Mitchell 2002, p216)[23].

In sum, the large investments in hydraulic infrastructure on the main and Blue Nile for *national* development during 1960s had not only closed the river basin but subsequently prefigured the profound rearrangement of *international* relations of debt and trade. From the 1970s the Nile flow downstream of Aswan showed little variation, with monthly

---

[23] Over the years the financial problem this created for the majority of the Egyptian population were compounded by ´environmental´ problems. Soon after the intensification of irrigation that followed the construction of the Aswan High Dam 80% of the plots in the delta faced problems with high ground water levels and 25% of the plots showed increasing salinity (Aboukhaled et al 1975). Since the construction of the Aswan High Dam the shoreline of some parts of the delta has receded at an average rate of almost 100 m per year due to coastal erosion (Hereher 2011).

discharges between 3 - 6 BCM (Figure 2.5). Whereas the yearly outflow to sea and lakes had varied between some 85 BCM/year and 40 BCM/year before 1950 (Hurst 1957), by 1988 the Damietta and Rosetta Branches had been dammed and the amount of water pumped to the sea was reduced to a saline flow of around 14 BCM/year (MoI Egypt, UNDP, IBRD 1981) (Figure 2.6). Claims by the governments that it was due to their careful operational management of these reservoirs which created food security were a myth (Allan 2001). Not only had the projects failed to deliver on their promise of welfare for the population within its national boundaries, it left the treasuries of Egypt and Sudan with massive public debts and millions of Egyptian and Sudanese smallholders dependent on US fertilizers and grains.

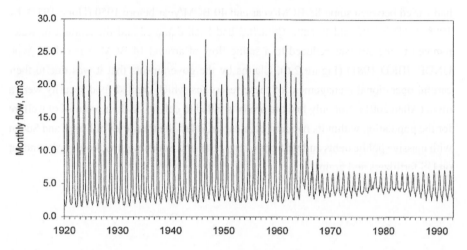

*Figure 2.5: Monthly Nile flows downstream of Aswan between 1920 and 1992- Note the regulating effect of the construction of the Aswan High Dam between 1960 and 1970. (Source: Sutcliffe and Parks 1999, p157)*

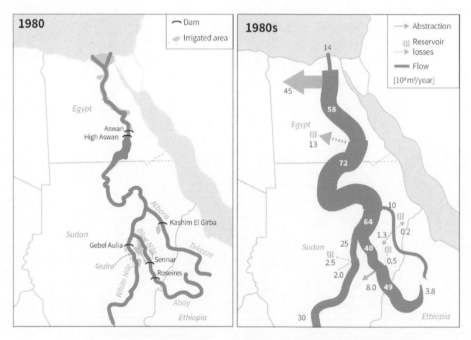

*Figure 2.6: Left: Dams after the second wave of dam construction on the Eastern Nile (1960-1970) Right: an estimation of distribution of the flow, abstractions and reservoir evaporation after the second wave of large scale hydraulic construction in an 'average' hydrological year.*

## 2.4    Hydraulic    (re)construction    for    green    growth    1973-2017 anticipating the future in an age of uncertainty

Once the physical limits to intensive agriculture for debt recovery became apparent in the closing river basin, the IMF and the World Bank shifted their calculations of water security. They mobilized the market to redress failures of earlier development efforts to take into account the value of the environment and the interests of future generations. From the 1980s, water security on the Nile was no longer merely a matter of harnessing and distributing supply of river water.  Water bureaucrats worked with World Bank experts on 'demand management' and 'cost recovery' to ensure that water would be assigned to more productive use(r)s (MoI Egypt et al. 1981, Abu-Zeid 2001, World Bank 2007, IMF 2014). The projects they supported to increase field level irrigation efficiency and re-use of drainage water would not reduce the consumption of Nile water. Whereas the second wave of dam construction had closed the river basin, the third wave would legitimize the *reallocation* of water committed by earlier projects to more efficient ventures that would protect the global population from increasing environmental and financial shocks.

Soon after the fall of the ´socialist´ regime of Ethiopia in 1991 the World Bank started to highlight how basin level cooperation could free up water for increasing production of crop and energy of the Eastern Nile region (World Bank 1994). Costs for new projects could be recovered, was the logic, from the net gains to be derived from increased hydropower production, more efficient irrigation and improved cooperation between the countries in the distribution of benefits along the river. For the Ethiopian government this discourse provided an entry point for accessing funds to put what it regarded as 'its' still 'untapped' water resources to productive use. For the Egyptian and Sudanese governments the discourse provided a way to justify a new round of mega-irrigation projects, while the promise of shared decision making enabled them to be involved in decisions over dam construction on the upper Nile. Building on these shared interests, the World Bank supported the launch of the Nile Basin Initiative by the ten Nile countries. In a parallel track, the United Nations Development Programme (UNDP) supported new negotiations over a Nile agreement that would include all riparian countries. Foreign consultants were engaged to support the preparation of projects in improving flood preparedness, increasing power trade and watershed management for international funding. By far the most prestigious project was the Joint Multipurpose Project (JMP) that would assist "the three Eastern Nile countries in the identification of a joint package of investments" (NBI-ENTRO 2013, p3) including a cascade of dams on the Blue Nile which would generate cheap and green electricity for the region and foreign exchange for Ethiopia. At the core of the project was the "spirit of cooperative development for maximum benefits and reduced risk" (NBI-ENTRO 2013, p3). Yet while the World Bank supported project seemed to extend Western influence over governance of the river through the diplomacy of benefit sharing, the frictions it was hiding were growing ever wider. In 2009, more than ten years after negotiations about a new Nile agreement had started, the JMP was still not forthcoming. With tensions over the project rising, Egypt's new water minister Mohamed Nasr Eldin Allam pulled out, after which negotiations over a new Nile agreement broke down as well. More importantly, the dominant role of Western credit organisations (the IMF, the World Bank and Western 'donor' governments) as financiers and mediators of dam construction in the South started to crumble.[24]

By the turn of the new millennium the liberal focus on good governance (i.e. fiscal austerity) that characterized most water development projects of the 1980s and 1990s had made way for a new burst of Asian and Arab investments. No longer was the World Bank in a position to condition upstream development on consent by all countries involved through integrated planning (Cascão 2009, Nicol and Cascão 2011, Yihdego et al. 2016, Cascão and Nicol 2016a). The deployment of Arab and Chinese credit, equipment and expertise in hydraulic construction went through the roof. Between 2006 and 2016 four large dams (Tekezze, Merawi, Rumela Burdana and GERD) and three major river diversions (Toshka, Salam canal, Tana Beles) were constructed on the Eastern Nile (Figure 2.7). Below I move downward along the river to explore how this new wave of hydraulic construction resonates with earlier projects along the river.

---

[24] As highlighted by Molle et al. (2006), mounting salinity levels and other environmental problems and limited returns to investments of postcolonial development had reduced support of international finance organisations for new large scale hydraulic works. There was another reason for the reducing World Bank investment in hydraulic construction; since the crisis of the 1970s the western creditors of the Bank had directed money elsewhere to reduce the rising costs of their own industrial production. Outsourcing manufacturing to Asia, US and European governments had themselves become the world's greatest debtors. As a continuing expansion of debt was required to keep major western companies going, the credibility of investments itself became the product of intense speculation. In need of raw materials, and prefiguring bursts of ballooning derivative markets, Arab investors and Asian ExIm banks and construction companies shifted their investments to African land, minerals, and infrastructure. This was the context in which food prices began to rise and showed extreme fluctuations in the new millennium. Between 2000 and 2012 trade between Africa and China grew from 10 billion/a to 200 billion/a, around double the trade between Africa and the United States (People's Republic of China 2013). In this new reality the World Bank – undisputed leader of financing dams in the global South between 1960 and 1990 - lost its role as a major player in dam construction on the Nile (Matthews et al. 2012, Verhoeven 2015).

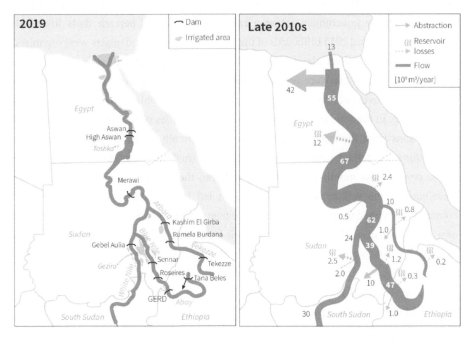

*Figure 2.7: Left: Dams on the Eastern Nile after the third wave of large scale hydraulic construction (2000-today) Right: an estimation of distribution of the flow, abstractions and reservoir evaporation with 2019 hydraulic infrastructure in an 'average' hydrological year.*

## Limits to Ethiopia's sweet hydro-electric renaissance

Chinese, Indian and Italian contractors and engineers enabled the start of 'Ethiopia's' hydraulic mission. The first large dam on the Tekezze tributary of the Nile was completed with Chinese investment in 2009. In 2010 a tunnel for hydropower generation was opened between Lake Tana and the Beles Tributary of the Blue Nile with Italian financial support and expertise. The real breakthrough followed in 2011. Two years after Egypt had reneged on the JMP project, three months after the eruption of Egypt's Arab Spring, and while Norwegian consultants were still calculating efficiencies of the JMP project, Ethiopia's Prime Minister Meles Zenawi announced that the construction of the Grand Ethiopia Renaissance Dam (GERD) had started.

During the inaugural speech the dam, with a reservoir twice the size of the one planned at the same location in the JMP, was presented as the centrepiece of Ethiopia's Grand Transformation Plan (GTP) (Zenawi 2011). Apart from hydroelectric investment, the plan presented large scale state led irrigation of sugar/ bio-ethanol production and large leases of land for irrigated farming to private investors as pillars for Ethiopia's transformation to a market based green industrial economy (GoE 2010). To ensure the buy-in amongst the population the government set up a campaign that would pervade daily life of all Ethiopians. From mid 2011 billboards of the GERD appeared, fund raisers were organised and agricultural trainings about the transformation from agricultural led industry to industrial led agriculture were organised around the country. Soon after, the campaign took more controversial forms: government employees were obliged to receive one month of their salary in the form of dam bonds. Private companies were forced to collect equal contributions from their employees[25]. And with agricultural development committees contributing and school children expected to make small donations in school, about the whole country was mobilized to literally buy into the project[26]. Apart from monetary contributions, the soil conservation campaign led by agricultural offices around the country was stepped up and 'requested' more than 10 million people to invest 20-40 days of work to prevent soil from eroding from their lands and 'silting the dam, or being wasted to Egypt'[27]. Those opposing to contribute or being critical of the dam were systematically labelled as 'against development'. Taxi drivers and shop owners who did not contribute on time were denied licences to do their business[28]. People in the countryside were threatened with lands being taken away[29].

The phrase of 'Ethiopia finally utilizing its right to harness its resources for development' was ingrained in a country-wide discourse of local institutions concerned with education, health and agriculture[30]. With one in every five members of development groups reporting on progress and 'abnormalities' in development to higher authorities, and the ruling coalition winning all seats in parliament in the 2015 elections, the 'developmental state' (Lefort 2012) seemed to ever tighten its grip. Yet within a decade after launching GTP and committing more than 10 billion USD to dam and sugar plantation construction (Matthews et al. 2013, Bloomberg 2015, Kamski 2016), doubts over the Grand Transformation have been painfully exposed. In July 2018 the newly appointed Prime

---

[25] Conversations with investors in Addis Ababa and Gumera 2011.

[26] Observations in East-Gojjam, 2012.

[27] As the Ethiopian news agency put it: "Farmers and pastoralists have been contributing about 79 billion birr [2.6 billion US dollar] worth labor annually by carrying out soil conservation work for about 30 days in a year to make the dam serve for a long period." (ENA 2018).

[28] Conversations with taxi- and tuktuk drivers in Addis Ababa and Arba Minch 2014 and shop owners in Arba Minch 2014.

[29] Observations in East-Gojjam 2011-2012.

[30] Observations at schools and healthcare and agricultural development trainings in East Gojjam, Addis Ababa and Arba Minch Zuria 2011-2014.

Minister Abiy informed parliament that 14.7 billion USD debt[31] had been incurred by state corporations in charge of megaprojects and their consistent reports of delay and reduced income had forced the government to refrain from signing off any new stated funded mega project (Abiy 2018). Perhaps one of the most dramatic instances of failed state investment is the billion US$ Beles sugar scheme, for which tens of thousands of farmers were removed. Sugarcane was planted since 2013. With reportedly 90% of the money spent the sugar refineries have for years been reported as unfinished and no sugar had been processed by 2018 (Fantini et al. 2018). The costs of the mega projects, which contributed to double digit growth and a booming Bole district of Addis Ababa, are borne by the Ethiopian population. With inflation hoovering between 25% in 2011-2013 and around 10% between 2014 and 2016  (World Bank 2017) and wages failing to keep up (Bachewe and Headey 2017) dam bond owners and Ethiopian consumers have been paying the price for massive speculation in land, hydro-electricity and sugar production. Discontent about rising prices, obligatory contributions to public works, political repression of opponents and ethno-political allocations of land to investors [32] came to a head in October 2016, when violence burst out around the nation. In the protests hundreds were killed, numerous foreign and domestic investment companies demolished and part of the Beles sugar plantation was set on fire (ESAT 2016, Abbink 2017).

A year and a half after the new prime minister took office, many political prisoners have been released and new state funded megaprojects were put on hold. Yet perhaps more significant reforms are taking place outside the capital, where millions have started to tap into the headwaters of the Blue Nile themselves. Responding to decreasing land holdings and rising food prices, they diverted streams and mobilized motor pumps to increase agricultural production and its reliability. Sometimes building on century old religious institutions and irrigation practices, and often supported by the government extension programme and NGOs, irrigation development of small entrepreneurs is rapidly rising (Awulachew et al. 2005, Dessalegn and Merrey 2015, Eguavoen et al. 2012). While hundreds of thousands of small hydraulic projects offer new opportunities and challenges

---

[31] This debt of 400 Billion Birr is well above Ethiopia's 347 billion birr government budget for 2018-2019 and does not include the 10 billion birr collected by bonds (Ethiopian Herald 2018).

[32] Of the 2.2 million ha of farm land - a large share of which located in the Nile basin - handed out to large private investors only some 15 percent has been used by them for agricultural production (Keeley et al 2014). Most of the concession holders are Ethiopians and diaspora who are widely perceived to be selected for their affinity and loyalty to the regime rather than on their ability and willingness to invest. Many have invested little and met with growing ethnic tensions, eventually resulting in the cancelling and in the cancelling and revoking of many of the leases issues (Dessalegn Rahmato 2014).

in water distribution along Blue Nile tributaries, water experts do not expect that the diversion of Blue Nile water after GERD in Ethiopia will exceed 3-4 BCM/a (Goor et al. 2010, Digna et al. 2016). While the GERD reservoir will evaporate some 1.6 BCM every year, its location at the end of the Blue Nile gorge and near the Sudanese border limit the potential for large scale irrigation development in Ethiopia.[33] Abstractions for small scale irrigation, on the other hand, are expected to be largely offset by increased drainage of high value irrigated plots in the wetter parts of the highlands (Tekleab et al. 2014). To better understand how GERD might rearrange the distribution of Blue Nile flows I turn to the plains along the Nile in Sudan, where the regulation of the flow by the dam increases possibilities for diversions of water.

## Dam boom in Sudan investment and gentrification of irrigation in Sudan 2000-2017

Despite, or rather because, of its long history of cotton production it is surprising that irrigation development has not been a priority of the masterminds of Sudan's new hydraulic megastructures. When Brigadier Omer El Beshir and his officers took control over a heavily indebted Sudan on the brink of another civil war in 1989, he was careful to prevent a backlash from elites with interests in cotton irrigation. With a constituency in Islamic banking, the regime pursued a reorientation of international cooperation towards the Middle East in the name of 'Salvation' and political Islam (El-Battahani and Woodward 2013). Yet as the Sudanese government defaulted on debt obligations and Western countries pressured Gulf States to refrain from investing in Sudan, the country became increasingly isolated.

Right before a trade embargo was imposed by the US and assets of the government were blocked, it found a new credible ally: China. In 1996 the Chinese National Petroleum Corporation signed a breakthrough deal to construct a 1600 km oil pipeline from South

---

[33] While it is certainly true that the rugged terrain limits Ethiopia's potential for irrigation development along the Nile, it should be noted that Ethiopian government and USBR 1964 identified over 400,000 ha suitable for irrigation in the Blue Nile basin. Part of this area - e.g. at Ribb, Megech, Beles - is now being equipped with large scale irrigation infrastructure. With the recent announcement of downscaling of some of the large scale sugar development projects along the Nile (ESC 2019), we estimate that reduction of the Nile flow due to Ethiopian consumptive use from reservoirs and irrigation along the Blue Nile will remain limited to 1.6 BCM/a (approx. 1,080 mm/year x 1500 km$^2$) and 2.0 BCM/a (<200,000 ha x 1,250 mm/year) respectively in the coming decade.

Sudan to Port Sudan and a 2 billion USD oil refinery (Patey 2007). Soon after oil money poured in, the Sudanese government contracted the Chinese ExIm Bank, China Development Bank and the Construction Bank of China to work on dams at Merawi (completed 2009) and Roseires (heightening completed in 2012). The engineers of the Ministry of Water Resources and Irrigation were never involved. Instead a Dam Implementation Unit (DIU) was set up directly under the president's office. Between 1999 and 2013 its security agents and administrators worked with Australian and German consultants and Chinese and Sudanese banks and contractors to plow back more than 10 billion USD of oil rents in dams and 'side projects' (Verhoeven 2015, Abd Elkreem 2018). That the dam programme served not only to construct dams but also to cement a new loyal network of investors who engaged in highly speculative projects is a public secret. As Verhoeven succinctly put it, the DIU spent part of the billions for Merawi on "roads to nowhere, bridges for medical conditions hardly prevalent in the River Nile state and … an international airport that stands largely empty" (Verhoeven 2015, p210-211). The government envisaged that these companies would work with investors from Qatar, Jordan, Saudi Arabia and UAE who leased hundreds of thousands of hectares to engage in a new era of irrigation development (Woertz 2013, MoA 2008).

Yet the economy stalled just as suddenly as it had been kick started. When South Sudan seceded in 2011 and oil rents dried up, contradictions with earlier projects of water security came to a head. While the booming upper class in Khartoum increased its access to electricity, the states in which dams were constructed remained largely disconnected from the electricity grid (ICG 2013, p7). Until then the water supply to Sudan's colonial irrigation schemes had retained priority in water allocation under the Ministry of Irrigation. Yet with the completion of Merawi and the dismantling of DIU the balance started to shift. DIU security officers took over the main building of the irrigation ministry and dubbed it the Ministry of Water Resources, Irrigation and Electricity. Not only did the ever increasing volume of water consumed by the Gezira scheme limit hydropower production of the new major power plant at Merawi. The construction of the Merawi dam and the Roseires heightening had increased reservoir evaporation by some 3 BCM/a, pushing Sudan's water consumption close to the limit it agreed with Egypt in 1959. At the same time, the announcement of GERD changed Sudan's political possibilities for hydraulic development on the Nile. Positioned between two acclaimed hydraulic empires with different views on distributing the Nile's water, Sudan took center stage in trilateral negotiations to fulfil its own hydraulic aspirations (Cascão & Nicol 2016b). Yet while railways, ginneries and the agricultural service department of the Gezira were sold and its bureaucracy dismantled to make place for a new era of efficient irrigation, since 2005 no more than 120,000 ha of newly leased lands have been put under irrigation.[34]

Hesitant to invest in expensive irrigation infrastructure for export agriculture in a climate of import restrictions and protests against rising food prices, investors instead showed an interest to use old channels and re-invest in 'inefficient' (post)colonial irrigation schemes. Investors from China and the Gulf work with local investors to move capital into old colonial irrigation schemes and pursue new technology-intensive and labour-extensive ways of cultivating the Gezira (SMC 2017, Woertz 2013). Yet as we will see in chapter 4, the existing infrastructures and institutions are not simply put aside. On the contrary – despite all talk of disintegration, inefficiency and misery that surrounds the Gezira – old cotton infrastructure, bureaucracies and labour camps of the Gezira now form the bases for new ways of producing for thriving local markets. Tenants and workers – supported by retired irrigation engineers of the old bureaucracy and working with local and foreign investors – have resisted the takeover of land (Sudan Tribune 2010) and expanded their control over the majority of 25,000 hydraulic structures that distribute water over the gigantic Gezira canal grid. Managing to control most of the 9 BCM/a of water diverted to the Gezira with which I opened this chapter, they grow oil seeds, sorghum and vegetables eaten by some 4 million people in Sudan.

## Limits to 'Water conservation' through land reallocation and reclamation in Egypt

The above described upstream developments have not withheld recent governments of Egypt to announce new monumental plans for reclaiming the desert. On the contrary, just like the government of Sudan, they used a discourse of scarcity to justify public support for new 'more efficient' irrigation development. In order to 'conserve' water and increase production, so the government logic goes since the 1990s, new investment models and irrigation technologies are needed to move beyond the inefficient irrigation of small and irregular plots of the delta. In 1997 president Mubarak announced the construction of new mega-projects to quadruple Egypt's liveable area through creating a 'parallel Nile River' with highly efficient irrigation. Twenty years later, and after 600 million public USD invested in a pumping station and main canal to divert water from the Nile at Lake Nasser, only 5,000 ha out of 540,000 ha of the Toshka project have been developed.[35] Both the

---

[34] Most of the newly irrigated areas is owned by Sudanese investors. The area extent is based on Google Earth analysis 2005-2015 by the author, July 2018.

[35] Google earth analysis by author based on Landsat image 10-20-2014 on Google Earth Pro

few foreign entrepreneurs who did invest in the Southern desert and international organizations have blamed its remoteness from markets and challenging investment climate in Egypt (Sowers 2011). Rather than reflecting on rising public discontent and poverty in the Egyptian countryside, the present leadership of Egypt mobilizes their critique of investors to justify new and 'better' land reclamation projects. Promising to improve market connections and irrigation technology, the government in 2014 announced that 1.7 million hectares of land will be reclaimed from the desert to facilitate a new round of investments (ESIS 2014).

While the expansion of irrigation in the southern desert remains limited until today,[36] the impacts of land reforms and new expansion projects in and around the delta by a similar logic of liberalization for conservation have been all the bigger. Since the implementation of the 1992 land law the land price skyrocketed, and more than a million smallholders have been forced to give up their plots (Saad 2002). A small number of family-owned agribusinesses have used the liberalization of the economy to vastly expand their positions on Egyptian markets of meat and processed food products. Large investors have implemented vast projects of the official and unofficial reclamation at the fringes of the delta (Figure 2.8) (Ismael 2009). Receiving tax breaks, land deals, international funding and water diversions justified in the name of efficient production, the one percent largest land holders of the country took control over a quarter of Egypt's agricultural land (El Nour 2015).[37] Yet as water and land are ever more contested and poverty rates are rising to almost 30% (UNICEF 2017), the limits to increasing the reallocation of water to large agribusinesses in the Delta come into view.

---

[36] The largest expansion is at East Oweinad, with some 60,000 ha currently irrigated (Google Earth 2018) by pivot irrigation tapping into yet another new 'frontier' of fossil groundwater (ESIS 2015).

[37] This is not simply a story of the rise of finance capital at the expense of the state. With managers of business empires appointed in Mubarak's cabinet of ministers, cooperation between state and investors was further intensified (Ismael 2009, Dixon 2014).

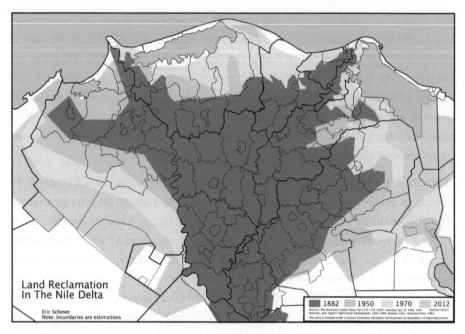

*Figure 2.8: Land reclamation in the Nile Delta 1882-2012. (Source: Schewe (2012)*

Lowering of canal levels and reuse of drainage water to enable expansion of the cultivated area push up salinity levels in the Delta (Zhu et al. 1996) and limit possibilities for further savings of irrigation water (Ritzema 2009, Molle et al. 2018). Crop water deficits and yield reductions due to increasing salinization have been reported in recent studies of water management practices (Ghazouani et al. 2014, Kubota et al. 2017).

With water deficits rising and inequality in land holdings higher than during the revolution of the 1950s, rural protests have burst into the open from the late 1990s and led the way for to revolution in 2011 (El Nour 2015). While policies of agricultural growth for export have remained after the fall of Mubarak, important changes are emerging in the organization of land and water. Faced with increasing difficulties to get hold of water for expansion of farming on new lands, the rapid territorial expansion of agribusinesses in Egypt has been halted as they have started to offshore their agricultural operations to other countries[38] (Dixon 2014, Sandström 2016). At the same time farmers in the Delta, who face ever more limited and saline water flows, are adjusting their cropping patterns and production practices. In the Western Delta for example, farmers

---

[38] See Hanna (2015) and Dixon (2014), who report that some agribusinesses have shifted farm operations to Sudan. As pointed out above, expansion of the irrigated area due to new foreign investments in Sudan has been limited.

have relocated and clustered the cultivation of rice along drains to reduce seepage and add drainage water when needed (Ghazouani 2014, p18-19). In the Salam Canal project area east of the Delta, farmers have taken control over the water ponds that were constructed to improve the soil. As the saline ponds were never reclaimed for the intended purpose of the project - high tech cultivation of vegetables for Europe – farmers now use the ponds for aquaculture and rice cultivation. Mixing 'fresh' water with salty water from drains, they supply their plots for new investments of intensive tilapia, mullet and rice production (Barnes 2014, Soliman and Yacout 2016). While these new 'models' for agriculture might seem inefficient in the eyes of economists and other experts who suggest that that production of drip irrigated export crops would enable to buy more food than currently produced (Breisinger et al. 2012, Osman et al. 2016), for many in the delta these alternatives are more promising than engaging in production for volatile and heavily subsidized European markets.

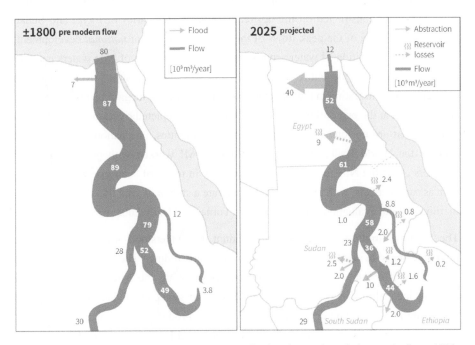

Figure 2.9: Two flow schematics of the Eastern Nile showing projected changes in flows 1800-2025

## 2.5 Conclusion

This chapter reconstructed the Nile's contradictory epoch of the hydraulic mission, in which the securitization of the river rendered access to land and water ever more uncertain for millions of people along the river. While the construction of large dams and mega-irrigation schemes have enabled a vast expansion of crop and electricity production, they were accompanied by violent appropriations of land, water, labour and public money, which have raised inequalities and deepened conflicts throughout the Eastern Nile basin. Three waves of investment in large scale hydraulic infrastructure over the last 200 years have each shifted the definition of, and limits to, water security along the river. The first wave of construction of large barrages and dams – starting in the 19th century to enable the expansion of cotton cultivation – led to the depletion of almost the entire base flow of the Nile. The concerns that arose in regards to safeguarding this base flow from upstream consumption invoked the invention and conquest of the *Nile Basin* territory. A second wave of hydraulic construction took place from the 1950s to the 1970s. The postcolonial dams that were constructed on the upstream national boundaries of Egypt and Sudan enabled the storage of the peak flows of the Nile in the name of *national* development. With this, the definition of water security and the consumption of Nile waters came to encompass (almost) the entire river flow. Yet this did not stop governments and companies from investing in the contemporary, third, wave of mega-projects on the Nile, which started at the end of the 20th century. Faced with depleted rivers and old irrigation schemes that are no longer producing export commodities, international organisations such as the World Bank, the IMF, FAO and IWMI worked with governments and investors along the Nile to redefine water security in terms of "ecosystems [which] are seen as capital assets, with the potential to generate a stream of vital life-support services meriting careful evaluation and investment" (Turner and Daily 2008, p25).

The shifting limits to water security cannot be understood separately from the rearrangements of land, water and labour they are hiding. The exponential rise in production of cotton, sugar, fodder, and electricity along the Nile are mortgaged on growing inequalities in welfare, new racialized and gendered divisions of work, and staggering rises in public debt. As capital intensive production for export concentrated large tracts land along the Nile in the hands of a few, a large share of the Egyptian and Sudanese people have become dependent on increasingly expensive wheat imported from outside Africa. With land and water being increasingly committed to large scale agriculture the outflow of the Nile has been reduced by some 85% (Figure 2.9) with the remainder unsuitable for consumption but required to halt further intrusion of salts in the Delta.

While large scale hydraulic construction along the Nile is in full swing in the name of increasing efficiencies, green growth and resilience, the sustainability of these new investments has become increasingly uncertain. As the infrastructures and institutions of the hydraulic mission made the river more sclerotic, the illusion of its premise of control over the river has become ever more obvious. Almost a quarter of the Nile water is now evaporating from human reservoirs built in the name of water security. With land and water ever more contested, investments in new hydraulic infrastructure have become increasingly speculative in the sense that (a) large tracts of lands are leased by foreign investors but remain undeveloped, and (b) investments in new large dams in irrigation that do materialize are increasingly subscribed by public debt rather than production. The rising cost of living this created along the Nile have not only sparked massive protests; With the reducing prospects of profit from new mega-investments, new spaces are opening up for millions of people who were subsequently presented as backward, underdeveloped and lacking resilience, to carve out their own projects of water security. In the following chapters I will explore the making of some of these spaces by following some of these people. In this way I aim to locate and relate some of the contemporary struggles over hydraulic construction on the Nile.

**3**

# The political morphology of drainage – how gully formation links to state formation in the Choke Mountains of Ethiopia [39]

[39] This chapter is based on Smit, H., Rahel, M.K., Ahlers, R., Van der Zaag, P., The Political Morphology of Drainage—How Gully Formation Links to State Formation in the Choke Mountains of Ethiopia, World Development (98), pp 231-244, https://doi.org/10.1016/j.worlddev.2017.04.031

## 3.1 Introduction

In May 1998 an Ethiopian farmer refused to drain water from upstream plots over his land. He blocked the flow and drained the water down the slope along the boundary of his plot. As soon as the first heavy rains of the season fell a month later, a gully was created 300 metres further down the hill. In the 15 years that followed, this gully grew 230 metres long, 70 metres wide and 8 metres deep, eating away the plots of six households; and it continues to grow.

Land degradation in Ethiopia is often presented as the 'natural' outcome of a growing rural population that is not capable of conserving the soil (e.g. Shiferaw and Holden 1999, Osman and Sauerborn 2001, Hurni et al 2005).[40] Since the 1970s, government officers and donor agencies in Ethiopia have worked with the rural population on soil and water conservation and 'good land governance' (FAO 1986, MoA 2013). More than 30 years of soil erosion research in the highlands of Ethiopia has demonstrated the possibilities of a range of soil conservation techniques to reduce soil erosion (e.g. SCRP 2000, Gebremichael et al. 2005, Nyssen et al. 2007, Frankl et al. 2011, Taye et al. 2013). Yet, despite massive investments in soil conservation, erosion remains severe, especially in the humid parts of the highlands (Hurni et al. 2005, Monsieurs et al. 2015a).

This chapter analyses what is arguably the Nile's ever biggest hydraulic project. Its subject is not one of the gigantic dams currently under construction along the Nile, but something seemingly much more mundane: a few of the hundreds of thousands of terraces constructed every year by millions of Ethiopians who are recruited for some 20-40 days a year to construct terraces to protect the land from eroding. By documenting the making of the above described gully the chapter address a straightforward question: Why is soil erosion on the hill slope persistent despite decades of popular mobilization for soil conservation? To answer this question I draw on studies of political ecology which identify social relations of production and the nature of the state as key factors in explaining environmental transformation (Blaikie 1985, Blaikie and Brookfield 1987, Andersson et al. 2011). Here I build on the work by scholars who explored how the Ethiopian government mobilizes its 'developmental state' model to reinforce state power under the guises of democracy and technical packages of development (Lefort 2012). The agricultural extension service has received particular attention in this regard, as it makes up the densest state bureaucratic network in the Ethiopian countryside (Vaughan 2012, Planel 2014). Political analyses of what are widely presented as technical development

---

[40] For exceptions see Ståhl (1990) and Lanckriet et al. (2015).

packages provide valuable insights into how practices of land registration (Chinigò 2015), Green Revolution (Teferi 2012), decentralization (Emmenegger 2016, Chinigò 2014) and input provision (Planel 2014) have been instrumental in the expansion of state power. In particular, programmes of mass mobilization have been highlighted as vehicles to implement the 'developmental state' model, both in the countryside (Dessalegn Rahmato 2009, Segers et al. 2009, Emmenegger 2016) and the city (Di Nunzio 2014). The limited attention for the materialization of these programmes is remarkable: how has the 'developmental state' model shaped and been shaped by the distributions of people, land and water in the landscape? How these programmes materialise is made clear by developing a political morphology of drainage.

Section 2 elaborates upon what it means to employ a political morphology approach. In section 3 the approach is operationalised by analysing how socio-material relations of drainage are literally scoured into a hill slope of the Choke Mountains. The conclusion sheds light on the (limited) powers of Ethiopia's 'developmental state'.


## 3.2 Methodology: Towards a political morphology of landscape transformation

Soil erosion is a classical object of political ecology (Robbins 2012). In his path breaking work, *The Political Economy of Soil Erosion in Developing Countries,* Blaikie (1985) explored why land degradation and social marginalization often go hand in hand. His account was followed by a wealth of studies on "the political, social and economic content of seemingly physical and 'apolitical' measures" (Blaikie and Brookfield 1987, p xix) commonly put forward to curb environmental degradation (for Ethiopia e.g. Hoben 1995, Keeley and Scoones 2000, Segers et al. 2009, Chinigò 2015). While the co-production of societal values, environmental knowledge and the physical environment is often claimed as central in this literature, the morphology of the landscape often figures as a result of this production process but not as its constituent. In this way, the instrumentalist studies of soil erosion critiqued by Blaikie in the first place are replaced (or complemented at best) by disembodied accounts of environmental knowledge production and resource extraction. To overcome this divide, this chapter moves away from the epistemological search for an accurate representation of social or physical processes. Instead I zoom in on the ontological question of how the morphology of a hill slope comes into being (following Mol 2002) and how this process can be accounted for. This political morphology approach resonates with accounts that analyse how political power, technologies, and environmental knowledge are relationally formed in the distribution of flows of land and water (Gandy 2002, Van der Zaag 2003, Mollinga 2014, Meehan 2014, Barnes 2014).[41]

The focus in this chapter is not on morphology as an expression of cultural forms (Sauer 1925) or on the ideology of depicting morphology (Cosgrove and Daniels 1988) but on accounting for the practices through which the hill and its users interact and morph together, i.e. accounting for the morphodynamics of the landscape. This shift entails a transition from the analysis of nature as a resource subject to domination or construction by humans, to an understanding of the socio-ecological process through which nature is 'produced', i.e. continuously transformed – mediated by technology – through labour (Smith 1984, Mitchell 2012). Gender, race and class identities are not taken as drivers but rather as products of these very material and discursive practices (Haraway 1991). Scholars of subject formation use this insight to show how the implementation of projects, policies and rules may create new collective identities that are often aligned with the interests of powerful actors (Agrawal 2005, Robbins 2007, Li 2007). In this chapter I mobilize this insight to analyse how international policy makers, government agents and land users are constantly at work to uphold the idea of a 'farming community' of the Choke Mountains, although many on the hill spend most of their time outside the farming profession. By paying close attention to the practices that reshape the hill I account for the active roles of people and material flows in shaping their identities and forms.

## Case study area and data collected

The analysis draws from observations and interviews in Yeshat kebelle (the smallest administrative unit in Ethiopia) in the Choke Mountains (Figure 3.1) between 2009 and 2012[42]. Together with 23 other *kebelles*, Yeshat[43] *kebelle* is part of Sinan *woreda* (district) which currently has around 60,000 inhabitants. Yeshat *kebelle* consists of 10 *goths* – parishes in which people attend the same church or *idder* (religiously oriented institution

---

[41] In a broader sense the chapter can also be seen as a contribution to the wide range of efforts to 'rematerialize' human geography (see Bakker and Bridge 2006, Whatmore 2006, Kirsch 2015 for discussions of the 'material turn').

[42] Unless otherwise indicated, the information in this chapter is based on research by the author and Rahel Muche Kassa, Tefera Goshu, Tesfay Muluneh, Atsbha Bhrane, and Temesgen Tefera who worked with me on their PhD/MSc/MA research along the Jedeb River. Quotes in this chapter come from the interviews held by these researchers.

[43] Names of people and places have been changed throughout the dissertation, except for major cities, rivers, mountain ranges, irrigation schemes, presidents and ministers.

through which burials are organised and through which people are mobilized for communal activities such as bridge and path construction). The altitude of the *kebelle* varies between 2400 and 2700 meters above sea level and the average annual rainfall is around 1400 mm/a (Tekleab et al. 2014). In particular this chapter draws from: 1) observations of people's activities and the functioning of drainage and soil conservation

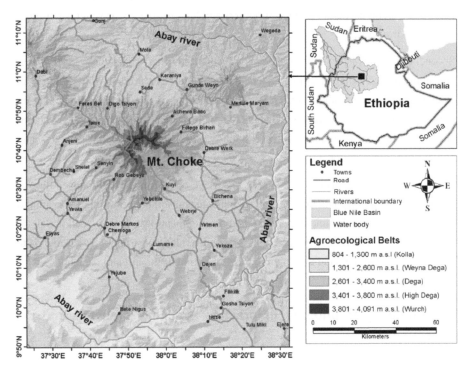

*Figure 3.1: Map of Mt. Choke and surroundings. The study area is located on the boundary between the agroecological belts indicated as Dega and Weyna Dega; Abay river = Upper Blue Nile river. (Source: Teferi 2015, p68)*

technologies on a hill slope that makes up the south of Michael goth (approx. 50 ha of hill slope, of which 38 ha are cultivated); 2) repeated conversations and interviews with members of the 14 households living on this hill slope and with 31 other households that were involved in cultivating its land or otherwise connected over a period of three years; 3) 24 samples of 2 m$^2$ of crops harvested from the hill slope in December 2010 and January 2011 to calculate grain yields and their variations along the slope; 4) an analysis of changes in the landscape based on discussions of aerial photographs of 1957 and 1982 and a satellite image of 2009 with people from Yeshat; 5) participation in meetings organised by government officials or *kebelle* leaders and an internship with extension agents responsible for the agricultural programme of the government; 6) rainfall data collected on the hill slope over a period of two years.

The next section explains the approach in three steps by first describing how social and physical objects relate in a particular event through which the landscape transforms (cf. Latour 2005). I follow a rain drop that fell during the storm of 9 July 2010 and drains over the hill to ground our morphology of drainage (section 3.1). This shows how drainage takes place along particular paths and borders and how people divert water according to particular strategies.

Second, I trace the history of the sociomaterial conditions that shaped these paths, borders and strategies (cf. Mitchell 2012), understanding the hill as a product of intertwined and changing relations of labour and geology. The people of Choke are not socially and physically positioned equally but caught up in historical and geographical relations embodied in physical boundaries, land holdings, and institutions such as sharecropping and oxen sharing (section 3.2).

Third, the chapter analyses how these historical conditions are actualized through people's contemporary practices related to drainage and soil conservation. By examining the organisation of a participatory watershed development programme (section 3.3), the drainage of a heavy rain storm (section 3.4), and terrace construction (section 3.5) the chapter seeks to understand the ongoing transformation of the hill and its people.

## 3.3. Political morphology in practice: tracking drainage and soil conservation

*Introducing drainage: following a drop along a hill slope in the Choke Mountains*

Intensive rainstorms at the start of the rainy season pose risks for cultivation in the Choke Mountains. Whereas government officers and academics often point out that the washing away of fertile soils will reduce soil fertility in the long run, the immediate concern of the ox-plough cultivators in the Choke Mountains is with flooding. To prevent storm waters washing away seeds, fertilizers or young plants, they use the traditional ox-drawn *maresha* ploughs to plough furrows that can drain their plots. By following a rain drop along the 1 km route where it hits the ground (A) to the point where it drains into the river (G), I discover that the practice of drainage is far from self-evident (see Figure 3.2).

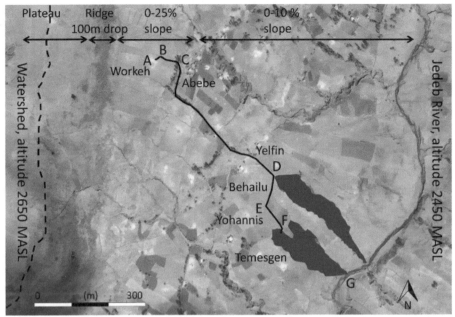

*Figure 3.2: Route of the water drop from the point where it hits the ground (A) to the point where it reaches the Jedeb River (G)*

*Note: The black areas indicate the gullies in 2010 – they are about 1 ha each. The hill slope studied in this chapter is characterized by a small plateau around the watershed and a steep ridge with bushes, rocks, and patches of grass above Workeh's plot. At the foot of this ridge starts a patchwork of cultivated plots, typically 0.1–0.5 ha each, draped down in strips towards the river. The slopes of the cultivated plots vary between 0% and 25%. The 100 m strip along the river is used as communal grazing land. (Source base map: Google Earth (c) 2009 Digital Globe)*

*From A to B*

On 9 July 2010 at around 5 pm, a raindrop hits Mr. Workeh's plot where he had planted wheat three weeks earlier (A – Figure 3.2). The drop detaches some soil particles and starts its journey down to the Jedeb River. Fifty millimetres of rain had fallen earlier that day and saturated the ploughed 30 cm soil layer. The compacted soil layer immediately below, the so-called plough-pan resulting from the many repeated earlier ploughings, made it difficult for the water to infiltrate. The drop flows quickly to the nearby furrow, which intercepts it and transports it to the plot boundary. Workeh had subtly increased the slope of the furrow along its way to accommodate more water as it flows down. He derives this knowledge from his long experience with draining the plot: heavy rain showers turn steep furrows into gullies. Furrows that are not steep enough overflow and flood the plot. Too many furrows cost energy to construct and use up part of the cultivated land. Too few furrows create harmful overflows that erode the soil. After travelling 15 metres through the furrow, the water reaches the plot boundary (B).

*From B to C*

Here, it joins the water from the deforested hill slope and from the neighbouring plot. The drop now moves along the boundary between the two plots straight down the slope. It speeds up. Every eight or so metres another furrow adds water to the drain. The concentrated downhill flow scours the drain and deepens it. To reduce the scouring, Workeh had covered the drain with stones. Despite his efforts, earlier flood events had already created a 1.5 metre deep gully over the last 10 meters of the boundary (C).

*From C to D*

The water flows 30 metres down a steep part of the hill slope with eucalyptus trees and quickly reaches Mr. Abebe's house and vegetable garden. Fifty metres further, the drop ends up in a protected drain. Abebe had managed to fill the 'man deep' gully along his plot by planting eucalyptus trees at both sides and constructing check dams every 20 metres. As the protected drain is too small to carry all the water down, it washes over a fifth of Abebe's plot and destroys half of the young maize plants on this part of the plot.

*From D to E*

Just before the water reaches the head of a 7-metre deep, 45-metre wide, 330-metre long gully (D), it hits stones that force it to make a 90-degree turn towards the south. It is here where a diversion was made in 1998 when the gully approached Ms. Yelfin's house. The sudden turn slows the flow and makes the coarse soil particles carried by the water deposit and clog the drain. The overflowing water washes Mr. Behailu's plot and buries his small wheat plants under a layer of fertile soil. The rest of the water and soil continue through the drain until it reaches the plot of Mr. Yohannis (E), who blocked the flow with stones that direct the water down the slope.

*From E to G*

The water drop moves along the plot boundaries until it drops into a gully (F) that takes it to the river (G). This gully had emerged a year after the diversion in 1998. Within two years, a 3-metre wide gully developed. From the riverside, it 'grew' up the hill eating into six plots along the drainage route. Since then, the gully has grown 230 metres long, 70 metres wide, and 8 metres deep (Figure 3.3).

*Figure 3.3: Mr. Temesgen ploughing with oxen next to the gully formed in Michael since 1998*
*Note: This is a picture of the southern gully in taken from point F in Figure 3.2. The Jedeb River flows between the slope on which Temesgen is ploughing and the slope at the upper part of the picture.*

## Historicizing drainage: on the (re)distribution of access to land and nutrients 1950–2010

The above account follows only one raindrop from Workeh's plot to the Jedeb River. Along the way, 25 more land users drain into the same route. Their relations are not merely charged by the risks of occasional overflowing. This section shows how the manoeuvres described in the previous section were configured by particular regimes of land distribution and rent appropriation that shaped the hill since imperial times.

The division of plots into strips over the hill (Figure 3.3) and the consequent drainage pattern have to be understood in the context of the land tenure system that evolved with ox-plough cultivation in Northern Ethiopia from the thirteenth century. The essence of the *rist* system is that people claim rights to a share of the land of their ancestors who are considered as original rightholders of the area (*wanna abbat*). In principle, the land was to be shared by male and female heirs. The people on the hill claim their rights as shares of a larger area held by the *rist* holder from whom they descended and do not attach rights to particular plots (Hoben 1973). Considerable ambiguity is created by the fact that heirs who do not inherit land after the death of a family member do retain rights to the land and can pass these on to their children. In practice, this often means that sons who were too young to cultivate and the offspring of daughters who often did not claim land at the time of inheritance claim land later.

Two elements of the *rist* system that persist are important for understanding land tenure and drainage of the hill slope. First, the multiple possibilities through which one can gain access to land make that the amount of land claimed by people is larger than the area of land available. Second, the ambiguity created by the multiple claims enable 'the authorities' to exercise power in resolving land and natural resources management disputes. As an elder stated when explaining his drainage route: *"The water flows the way the king wants it."* The outcomes of disputes over land and its drainage depended crucially on the contestants' abilities to convince the local court, which has recently been embedded in the government court system.

After the overthrow of the imperial government (1974), landlordism and the *rist* system were formally abolished and all land was proclaimed public property with the official aim to correct injustices of imperial land tenure (Desalegn Rahmato 2009). In practice this did not mean the end of the influence of the former landholding class. Mr. Eshetu, the *rist* holder of the hill slope, was elected as both chairman and leader of the militia (designated armed civilians who assist the *kebelle* administration to maintain law and order) of the Peasant Association (PA) in Yeshat *kebelle*.

Officials of the Dergue regime (that ruled Ethiopia from 1974 to 1991) in the *woreda* capital worked with Eshetu to re-distribute his own lands. Yeshat's land committee divided Eshetu's land between himself, his adopted son Mr. Temesgen, and the three tenants, Mr. Negus, Mr. Yohannis and Mr. Mersha, who had cultivated his land until then. Two women servants, Ms. Wolete and Ms. Ashene, received part of the grazing lands for cultivation.

Lands on the top part of the hill, which was partly covered by trees and partly leased out by the church for grazing, were distributed to other families. Between 1976 and 1982, the number of families on the hill slope doubled from 12 to 24. New families cleared acacia and juniper trees from steep slopes and around major drainage routes for construction and fuel. As many of these families could not increase their landholding until their parents died, they reduced fallowing of lands to intensify cultivation. This shift was enabled by the introduction of inorganic fertilizers that were heavily subsidized and provided on credit by the state.

Although, initially land distribution was welcomed by most people living on the hill slope, the Dergue government pushed through increasingly unpopular measures. The role of Peasant Association leaders significantly changed relations between people cultivating the hill. Mr. Eshetu was not only involved in the distribution of land but also became responsible for the collection of the grain quota. Between February and June every family had to sell roughly a third of the harvest at a little over half the market price. Mr. Fekadu, who had one of the upstream plots draining to the gully, was put in charge of recruitment of young men for the army and for mobilization for the yearly construction of terraces. He recruited one of the sons of Mr. Yohannis, who holds land at the downstream part of the slope, to fight revolutionaries in the north of the country.

At least four rounds of land distribution between 1975 and 1985 increased the cultivated area and increased disputes over drainage. A rising number of people refused to allow their neighbours' water to drain over their lands. By blocking the flow at the plot boundary, they refused them access to the main drains to the river. Fresh cracks appeared on new vertical drainage[44] routes along plot boundaries. Along one of the new drainage routes thus created, a 200-metre long, man-deep, and up to 15-metre wide gully had emerged by 1982 (Figure 3.4).

By the time the Dergue regime fell in 1991, fallowing had been almost abandoned. As more young families needed land, calls for land reform or new redistributions increased. The new Ethiopian People's Revolutionary Democratic Front (EPRDF, the coalition of ruling parties that was formed after the Dergue was overthrown) kept all land under state control and included the right to arable land for all adults in the countryside in the 1995 constitution (see Desalegn Rahmato, 2009). To correct the injustices of previous land distributions by the Dergue government, a new land redistribution was implemented in 1997 (Ege 2002). Many who held official positions in the Dergue regime, like Eshetu and Fekadu, were listed as *birocrat*, that is, associated with the ousted regime. Consequently, their landholdings were reduced to 1 ha, whereas 'normal' families (amongst whom the families of their sons, like Temesgen) were allowed to keep up to 3 hectares. Most of the land freed up was allocated to young families.

---

[44] Vertical drainage: drains follow the slope straight downhill, perpendicular to the contour line, thus have a steep gradient. Horizontal drainage: drains follow the contour line with a small gradient.

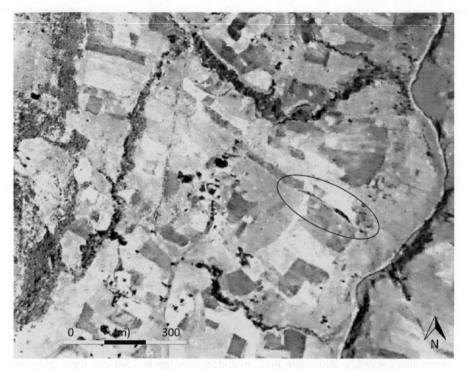

*Figure 3.4: Aerial photograph of the hill slope taken in 1982. The emerging gully encircled is the Northern gully in Figure 3.2. (Source: Ethiopian Mapping Agency)*

In the face of increasing competition over land amongst a growing population and aware of sensitivities of earlier redistributions, the government has refrained from distributing land to young households in Yeshat since 1997. Like elsewhere in the Ethiopian highlands (Ege 2015, Lefort 2012, Chinigò 2015) this has led to widespread landlessness amongst young households, postponed marriages, further intensification of cultivation and encroachment into grazing lands. Both the land under cultivation and cropping intensity have doubled since the 1950s, leading to a fourfold expansion of the cropped area.[45] On the 38 ha of land around the gullies, 122 of 130 plots were cultivated during the 2010 main cropping season (Figure 3.5). This intensification has come at a price. One cultivator on the hill explained that *"while soils wash away, only stones remain"*. Another stated that *"these stones were grown from my land like the crops I cultivate. These big stones in*

---

[45] We estimated the increase in the cultivated areas with the help of aerial photographs from 1957 and 1982 and a satellite image from 2009. I estimated the increase in cropping intensity based on interviews with people who have been cultivating in the area for a long time and on Dilnessa's (1971) study about cultivation on a hill slope at similar altitude, 20 km from Michael, in the 1960s.

*my land are creating difficulty for my oxen to plough even. The stones were not here before".* The plots halfway up the slope now have more than 20% of their land surface covered with stones.[46] Land users in Michael report that, after years of yield increases due to the introduction of new crop varieties and fertilizers,[47] yields are now falling because *"fallowing was abandoned and soils are being cultivated every year"* and *"the soil became addicted to fertilizer"*. Others stated that: *"If our lands have mouths to speak, they will tell us how much they are exhausted by ploughing all these years"* and *"lands are now deaf to hear our investment."* We found that average yields on the hill in 2010/2011 – reported as an average year[48] – are low at 1.0 ton per ha for *teff*,[49] 2.0 ton per ha for *adja dekel*,[50] 1.0 ton per hectare for oats (*ingedo*), and 1.1 ton per ha for white wheat.[51]

Farming has now become a costly business. Cultivated plots are ploughed on average five times before planting to reduce weeds and – as many in Yeshat say – to scrape fertile particles into the exhausted upper layer (Figure 3.6). Moreover, most crops no longer grow without fertilizers. Those who cannot afford fertilizers have turned to the cultivation of oats – which almost everyone in Yeshat now uses for making *injera*. As their oats yields are often around half the yields of people who grow the crop *with* fertilizers, many sharecrop out their lands in return for half the crop.

---

[46] Field observations and photograph analysis of 24 fields from which harvest samples were taken.
[47] Dilnessa's (1971) study reports yields of 0.5 ton/ha for both *teff* and barley – about half of current yields. Together with the fourfold increase in the cultivated area, this would mean around an eightfold increase in production since the 1960s.
[48] This was contrary to the south and east of the country where a food crisis would emerge over 2011 – which drove up grain prices.
[49] An annual grain whose flour is used to make sourdough-type flatbreads, known as *injera*.
[50] Latin name *Triticale*.
[51] The year 2010 is considered by the people of South Michael as normal in terms of yield, except for the white wheat crop, for which bad seeds were reported as the reason for a yield reduction of up to a third. These figures were on average 40% lower than those reported to the *woreda* office by the Development Agent.

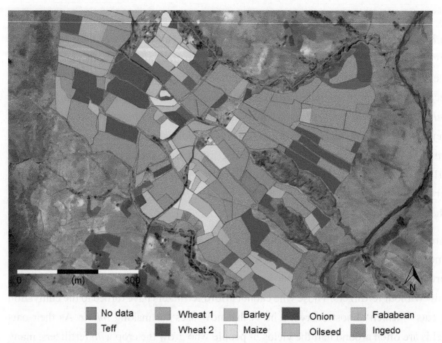

*Figure 3.5:. Crop map for the 38 ha cultivated area of Southern Michael, autumn 2010, wheat 1 = white wheat, wheat 2=* adja dekel, ingedo = *oats*

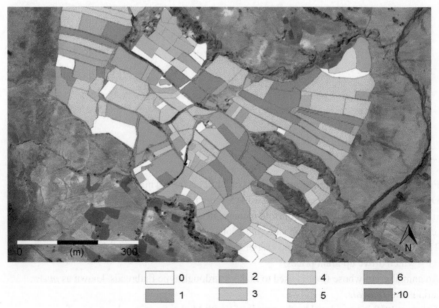

*Figure 3.6: Number of ploughings before planting the Summer 2010 crop for the 38 ha cultivated area of Southern Michael*

Agricultural intensification in Yeshat is paralleled with increasing social fragmentation. The involvement of some on the hill in redistribution of land and forced recruitment of others has strained cooperation between households. At the same time, the increasing scarcity of land sparks protracted struggles *within* households. Many young people from the hill have left the area. Yet, still half of the households in Michael *goth* have no access to land at all or only have a small garden plot. Some of them engage in sharecropping on the lands of old and absentee owners or landholding people who have no oxen to plough or money to cultivate. Almost all are involved in trade with the lowlands – 30 km downstream – that produce for the growing urban centres in Ethiopia.

The next sections explore how the superimposition of below-subsistence wages and petty commodity production on the cultivation of the socially and physically fragmented hill is transforming the shape of the hill, its people and the political economy. To elucidate how these changing relations of land, water, and labour are experienced by, and embodied in, different people and the eroding hill, I turn to the practices of party meetings (section 3.4), drainage (section 3.5), and soil conservation (section 3.6) between 2010 and 2012.

## 3.4 Contemporary practices of reshaping drainage 1: participatory watershed development

During the 12 months I lived in Yeshat as a neighbour of the Farmer Training Centre, the *kebelle* office, and the *kebelle* jail in 2010–2012, the boundaries between democracy and autocracy became increasingly unclear to me. The agricultural extension office launched a 'Community-based participatory watershed development programme' and recruited a 'development army' (2011) with the promise that these would bring democracy and development. Mr. Molla, the head of the natural resources management department of the Amhara regional state, explained to me that the focus is on democracy and participation and therefore *"the community will be involved in the planning, the implementation, and the monitoring and evaluation of the plan."* He continued that: *"[a] watershed committee is established in every 250–500 ha watershed that develops an integrated plan, which includes soil and water conservation, livestock development, crop production, and irrigation."* Yet, the participatory role that Molla had in mind for these 'representatives' did not fit easily with the top-down approach of the watershed programme, which included forced mobilisation of farmers. Molla explained:

> *"Something has to be done because fertile land is lost not only for them but also for the generations after them. Therefore, first we have to create awareness about this, but if they still do not want to comply after that, we have to enforce the measures to preserve the land for the coming generations.*

> *Therefore we have the land proclamation, which says that every farmer should conserve his land and provides a legal basis for this enforcement."*

As government officers at the *kebelle* level, the development agents (DAs) have a key role in implementing government and party policies. To the people of Yeshat, they are the face of the *mengist* – an Amharic word which "designates indifferently power as such, the ruling party, the state and all their agents or members" (Lefort 2012, p. 439). Mr. Yonas is the DA responsible for Michael *goth*. He was raised in the *woreda* capital 20 km down the river and was 21 years old when he moved to Yeshat. After obtaining his diploma in vocational training for agriculture, he subscribed to the EPRDF party to be able to join the agricultural extension office. Like the other DAs sent to the highland *kebelle* of Yeshat, he was inexperienced. Together with the other DAs and the newly appointed *kebelle* manager, he set out on his first important job: the expansion of the *kebelle* council from 60 to 300 people. With the local militia, they selected 280 out of around 900 Yeshat household heads (90% men) for a meeting on governance. The *woreda* officials had instructed the DAs to select landholding farmers, especially those who held land, were rich, or in leadership positions and could be an example for others. Whereas the main opposition leaders held responsible for the unrest during the 2005 elections were excluded, the landholding *birocrasi* were included in the new council. Together with the three DAs, 15 teachers, two health extension professionals, one policewoman, and one security guard who were posted in the *kebelle,* the selected people formed the 300 member council (300 *sew conferencegna).*

Land has been made the entry criterion for virtually all positions in the numerous development committees in Yeshat. 'Participation' and 'development' were used by the *kebelle* office to enrol the selected group in committees to which privileges were attached. During a 14-day meeting, for which all participants were paid 5 birr[52] per day, Yonas and his colleagues trained the council members about democracy and development and asked them to articulate the government's previous mistakes. In the end, all were enrolled as party members. In a second meeting, the *kebelle* parliament elected among its members committees for women affairs, youth affairs, societal issues, economic affairs, administration, and peace. For every *goth* a committee for watershed development was established. Yonas is responsible for giving technical advice to Michael *goth* and two other *goth*s in the *kebelle*. With the three committees, he develops almost identical lists of technical problems and solutions (Table 3.1). The committee members have little to choose as *"the* woreda *expects complete coverage – so we have to put these figures"*

---

[52] In December 2011, 1 US Dollar equalled 17 Ethiopian Birr

(Yonas). Once the plan is made, the committee does not meet again. Yonas complains that: *"we [the DAs] end up nagging the farmers. We don't have any other option really. We are like the messengers of the* woreda *BoA [Bureau of Agriculture] and administration office."*

In 2011 Yonas announced that development activities would become the responsibility of the newly established EPRDF political party cells (*hewas*) established for every *goth*. Its members are the people in the *kebelle* council. The leading role of the cell members in the implementation of the development agenda was formalized through their appointment as leaders of the 'development army'. Cell members were assigned responsibility for the 'performance' of five neighbouring households through the so called '1 to 5 system'. Every headman reports to the cell leader, who reports to the *kebelle* manager on a weekly basis. To ensure that the members attend the cell meetings, Yonas organizes these twice per month on Orthodox holidays, immediately before the religiously oriented local organization – the *idder* – meets. Despite this, participation is usually poor, with less than half of the members participating. The cell meetings follow a fixed agenda. First, the rules are listed and the fines in case of violation thereof. Yonas stresses that the *hewas* – and not the *idder* – is the sole authority for upholding the rules. Then members are invited to report on violations of these rules and other subversive activities. Third, Yonas informs the members about development activities that month. Yet, never in the cell meetings is there mention of the most pressing issue felt by the 620 families who are not represented in the *kebelle* council and committees: the expanding group of people who have no land or only a garden plot – many of whom engage in sharecropping.

Although attendance at cell meetings is often poor, its leaders and those who do attend engage in the performance. In return, they receive improved seeds, shovels or steel wire. More importantly, it gives them an advantage in their negotiations with the DA when the implementation of unpopular measures comes up. Government officials are also more likely to provide them support in times of disputes over land or drainage. Together, the members endorse fines for illegal encroachment on grazing land, cutting wood, and blocking drainage, thus producing a peculiar form of community-based participatory resources management. As the 'development' activities discussed focus exclusively on the farming domain, the council members model themselves after the image of 'the farmer'. As their actions align with party interests, they shape party rule.

*Table 3.1: Problems and solutions identified and targets set by the watershed management committee in Michael goth (2010)*

|   | Main problems identified | Solutions identified |
|---|---|---|
| 1 | Concerning plants | |
|   | - disease | *[Unreadable]* |
|   | - low fertility/acidity | Apply fertilizer/lime |
| 2 | Natural resources | |
|   | - water erosion | Cut-off drains, terraces, check dams |
|   | - deforestation | Plantation |
|   | - wild animals have left the area | |
| 3 | Grazing lands | |
|   | - incorrect grazing land management | Advise on grazing land management |
| 4 | Social and economic problems | |
|   | - no clean water available | Clear the sources every month |

**Targets**

| Item | Unit | Total from 2002–2005 | 2002* | 2003 | 2004 | 2005 |
|---|---|---|---|---|---|---|
| Terrace | Ha | 110** | 10 | 20 | 40 | 40 |
| Check dams | Km (gully) | 4 | 1 | 1 | 1 | 1 |
| Cut-off drain | Km | 4 | 1 | 1 | 1 | 1 |
| Outlet | Km | 4 | 1 | 1 | 1 | 1 |
| Tie ridging | Ha | 110** | 10 | 20 | 40 | 40 |
| Lime | Ha | 110** | | | | |

\* All years in this table are Ethiopian calendar years. The year 2002 in the Ethiopian calendar is 2009–2010 in the Gregorian calendar.

\*\* This concerns the whole land area cultivated in Michael *goth.*

The focus on participatory soil conservation – in its equation with development – serves to keep the bureaucracy together. Who can be against sustainable development? Whereas, for some, science provides a rationale for party rule, for Yonas it offers hope for recognition of his work. Science is not a naive excuse to justify working for a party that he resents. Yonas is proud to refer to the science of the *Guidelines for Community Based Participatory Watershed Management* (Lakew Desta et al. 2005) in explaining how he knows the world. For him, science provides a way to legitimize his authority and a possible avenue to BSc education. The walls of his office are full of tables and graphs showing progress. Often he refers to his education at the vocational training centre to emphasize the basis of science and policy in his arguments. *"If we don't implement this plan we all go down,"* Yonas explains. *"We can only escape this if we conserve our soils and transform the economy. Given its good rainfall and high altitude, experts decided that this* kebelle *will have to focus on potatoes. Others will provide maize that will be an input for an industry that will boost the economy."* It is in this technocratic agenda that Yonas, the party members, and the international organizations supporting soil conservation in the *woreda* find one another. Upward accountability of a story of a degrading hill to be rescued by modern science thus complements downward privileges of a selected group of farmers.

The expanding network of party members on the development committees in Yeshat celebrates its (s)elections as the fruits of participatory development. Land is the most important entry criterion for being selected and thus gaining access to state privileges. The focus on the farm in government development activities thus shapes both farmer identity and party rule. As the government presents its development activities as solutions for a homogeneous farming community, the different historical stakes within the state and the community are effectively silenced with the language of development. The performing of the farmer model not only shaped new solidarities between landholders and created a support base for the government, but also reinforced the tension between households with and without land. To see how the increasing differences in the community resonate with a hill slope that is far from homogeneous or static either, I turn to the practices of drainage and soil conservation.

## 3.5 Contemporary practices of reshaping drainage 2: Reconfiguring drainage routes

Whenever faced with a drainage flow from an upstream neighbour, a land user has two possibilities: 1) he allows the flow to pass over his plot or 2) he blocks the flow with stones to divert it down the hill along the plot boundary. If he accepts the water, it flows straight on and the water velocity remains the same. If the water is turned down the plot boundary, the water velocity – and thus the erosive power of the flow – increases with the slope gradient. Although steep drains can convey more water than flat drains, fast flowing water scours the waterway (Monsieurs et al. 2015b) and might, if not protected with stones or grasses, create rills.

Several factors have led to the increase in surface runoff in the rainy season over the past 50 years. First, as only a few trees remain on the hill, after a rain event less water is captured and water flows directly towards the cultivated area. Second, as more grazing land is now cultivated, the opportunities to dispose of water into safe drainage routes are limited (Teferi et al. 2013). Third, whereas more than half of the plots were fallowed in the 1950s, fallow is now hardly ever practised. Thus, the amount of cultivated land that needs to be drained has doubled. Fourth, extra furrows are ploughed into the fields to prevent the flooding of scarce land and the washing away of expensive fertilizers. Fifth the plough-pan hampers the infiltration of rainwater into the soil, increasing surface runoff during rainstorms (Temesgen et al. 2012).

The increase in surface runoff coincides with the growing unwillingness of many to drain their neighbours' water. As the approaching water is diverted straight down the border of their plots, the number of vertical drainage routes along the hill has increased (Figure 3.7). This is especially dangerous on the lower part of the slope, which is flatter and has loamy soils which saturate during the rainy season. This makes this part of the hill susceptible to the collapsing of slopes wherever rills are created. Whereas up the hill, where the slope is steep, protection with stones helps to stop gullies from growing, in the lower part such protection is often undercut and slopes easily collapse (see also Bayabil et al. 2010). In the 200-metre zone near the river, numerous gullies have thus appeared, resulting in the loss of land that was previously used for grazing and cultivation.

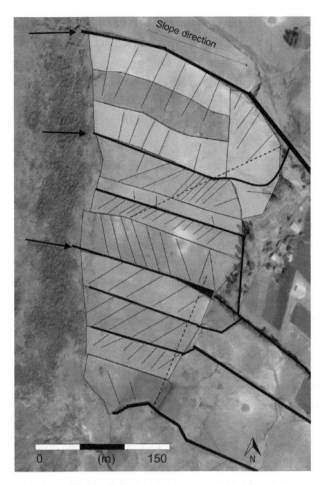

*Figure 3.7: Detail of pattern of furrows (thin lines) and drains (bold lines) upstream of Abebe's house (see Figure 3.2)*

*Note: The furrows of all but one plot end in 'vertical' drains that direct the water straight down the slope. The two dashed lines indicate the remains of two drains dating back to the Dergue period (1974–1991).*

However, the formation of drainage routes on the hill is not merely the result of more intensive drainage, saturating soils, and slumping gully walls. It depends on understandings of *who* are rightful cultivators of the land and *what* are 'natural' drainage routes. When the growing gully started to threaten the house of Yelfin's family, her deceased husband mobilized an acquaintance from the nearby agricultural office. During his visit, he ordered the people around the gully to dig two diversions to change the drainage route. When the extension officer left at the end of the day, they had not finished the southern diversion. Yohannis, embittered by the distribution (1997) of his son's land which he cultivated since his son was forcefully recruited and died in battle, refused to accept the prospect of losing more land. He stated: *"Maybe they think I am old, but I am not a fool. Why do they make their problem my problem? Therefore I blocked the flow [and turned it down]."* He blocked the flow and diverted it straight down towards the plot that he felt he had lost to Mr. Kassahun during the 1997 land distribution (the plot south of the emerging gully in Figure 3.4).

In the years that followed, the diverted flow washed away soil from a barren spot of grazing land downstream of Kassahun's plot. Soon, a new gully expanded along the entire boundary with his neighbour. Kassahun appealed to the DA every year but could not convince him to divert the flow. He explained: *"You must know we are not equal ... I was young and the women on the other land sharecropped out the land. Our plot is in between the plot of the son of the old* kebelle *leader and the plot of a militia."* Over the next 10 years, he would lose his whole plot. Kassahun understands well why Yohannis and his son Tadesse do not help him. All surrounding cultivators know the dangers of the saturated slope[53]. *"Why would [Yohannis] help to protect land that was taken away from him? Or why would [Tadesse] help to protect lands that are not his? "*

But Yohannis points out another reason for the growing gully: *"The new generation is poor at farming. ... They have no oxen but donkeys and engage in trade. ... They make no ditches or stone protections to protect the soil and crop ... Moreover they might claim the land. ... Every year [my son] moves the boundaries of his plot already. ... That's why I refuse to sharecrop my land to my sons."* His judgment does not only result from his frustration with the decreasing willingness of his sons to support their ageing father with daily work; it also resonates with the 'farmer image' reproduced again and again by landholders and extension agents.

For Tadesse, his father's inability and reluctance to provide him with more land and oxen makes it impossible to engage in farming. Not only does he lack land for cultivation, he is also barred from access to credit because he has no collateral. With soils increasingly requiring fertilizers, the price of which rose 40% in 2010 alone, access to credit is essential for farming. As Tadesse's access to the farming profession is thus severely constrained, for him the meaning of the hill slope has radically changed. Tadesse trades together with Kassahun, who explains how *"the donkey is now even more important than*

---

[53] Sadly and ironically, Yelfin's husband died in a gully less than 500 metres from the gully that emerged after his complaint about drainage.

*the oxen as we use it all year round for trading.*"[54] During his latest five-day trade expedition, he bought 24 bamboo baskets from the nearby highland market for 128 birr. The next day, he started his journey to the lowlands, a two-day walk from Michael. Here, he exchanged the baskets for 80 kg wheat and 15 kg maize. He sold the wheat in the *woreda* capital for 296 birr and took home the money and the maize. His wife, Ms. Anemo, sells some of the maize in small quantities in the nearby market. Moreover, she cultivates a small irrigated potato plot and engages in the production sale of *araki* (liquor). Whereas Tadesse's money is used to engage in new trading activities and the purchase of more grain when needed, the money Anemo earns is used to buy sugar, salt, and oil, and to meet other living expenses.

Failing to live up to the farmer ideal, young families in Yeshat have thus created a competing 'trader model'. Most young families in the *kebelle* complement farming of a small garden or sharecropping a grain plot with trade in baskets and lowland grains. Two years ago Kassahun, Tadesse and 33 other young traders[55] established their own *mehaber*, a kind of cooperative society, with its own saving scheme. They saved 4,500 birr and provide loans to individual members at 10% interest per month. The members accompany and support one another during trade trips to distant markets. Kassahun proudly tells me that this *mehaber* is now stronger than any other *mehaber* in the *kebelle*.

Changing relations of production in Yeshat are thus closely tied to the thriving lowland economy. In the lowlands the government had started leasing out previously uncultivated communal grazing lands[56] and providing cheap credits to state selected investors, cooperatives and development organisations as part of its new policy of commercial agriculture for development (EPRDF 2007) which was adopted after the disputed 2005 elections. On market days, most of the donkeys leave Yeshat empty to bring in maize which is now the cheapest grain available in highland markets. From September to December, these grains bridge the food gap until grains are harvested again in December–January.

Yet, as trading requires long distance walking and wages paid in commercial farming in the lowlands are below subsistence, Kassahun and Tadesse still aspire a career in farming. They are keen to sharecrop with their parents, especially because this supports their

---

[54] The 53 families in Michael *goth* own 58 oxen and 48 donkeys.

[55] Almost all men born in South Michael who were between 20 and 35 years of age in 2011, control less than 0.3 ha land and do not own a pair of oxen for ploughing.

[56] This does not happen without resistance: during a violent outburst in 2011, lowland cattle owners killed the oxen of investors plowing their grazing land and put their houses on fire.

claims to land. This leads to rising conflicts with their parents who are reluctant to sharecrop land with children who stay away from the land and church. When Yohannis decided not to sharecrop with his son in 2012, Tadesse hit his father with a stick during ploughing. Yohannis shouted "*I prefer to die today. You are coming to beat and kill me for land. Don't forget I am your father who suffered a lot to support you, starting from your childhood.*"

Yonas, the DA, manoeuvres carefully in the tense spaces thus created. Kassahun requested him to divert the drainage water from his eroding land every year, but landholders along alternative routes count on Yonas not to make any change. Along the new route two entire plots have been washed away and six other plots have been affected since 1998. Yet Yonas knows the dangers and sensitivities of diverting a drain. He fears becoming involved in any rapidly widening gully. Instead, he uses the widening gullies to justify his work on the government's soil conservation programme.

## 3.6 Contemporary practices of reshaping drainage 3: Turning terraces into drains[57]

Every year in January, the *woreda* officials order the yearly soil conservation campaign. Democracy is temporarily put aside as 'the community took no action and was under immediate threat.' Throughout Amhara regional state (20 million inhabitants), farmers are ordered to work for 40 days on soil conservation. The officials refer to the science of the Swiss-funded Soil Conservation Research Project (SCRP 2000) to justify this measure. "*The trick is,*" the natural resources management officer says "*to upscale the good practice of the SCRP programme.*" Under the SCRP programme, terraces were constructed by digging trenches and throwing the soil up the hill. Thus, local obstructions are created for water flowing down the hill. When the water flows down, the velocity drops, and silt deposits behind the bunds to form a terrace (Figure 3.8).[58]

The soil bunds thus created trap not only soil, but also water. During heavy rain storms this can lead to sudden accumulations that create breakages in the bund. When one bund breaks, this often creates a domino effect in which water flows further concentrate, and multiple bunds are broken. If terraces have not been firmly established and precisely laid

---

[57] The data on the terracing programme in 2011 were gathered with Bhrane (2012).
[58] In the manual for community-based participatory watershed management (Desta et al. 2005), the Swahili term *Fanja Juu* is used for this technology.

*Figure 3.8: Cross-section of a so-called fanja juu for terrace construction as specified in the Ministry of Agriculture and Rural Development guidelines by Lakaw Desta et al (2005, p76). Note: The figure suggests putting the excavated soil above the ditch to construct the bund and allow it to develop into a benched terrace.*

out along the contour, there are considerable risks of breakages and flooding. The SCRP project funded the construction of a medical clinic to compensate the people on the pilot sites for prohibition of grazing for three years to protect the new terraces and for frequent breakages in the two years after terrace construction.[59] Yet as Sinan *woreda* does not have the resources to compensate for losses, extension agents and land users have transformed the design to prevent harmful floods. The soil is thrown downhill from the ditch instead of uphill, thereby creating graded ditches that can drain the water accumulating behind the bund. Although the implicit idea is that the bund will strengthen and benching will take place in the years that follow,[60] terraces are constructed in such a way that benching never happens.

From the outset, it was clear that the terracing programme was far from popular. Militia are mobilized to ensure attendance at training events at least for the first few days (Figure 3.9). For example when Yohannis' youngest son, Demaiferam, returned from a trading trip at night, they held him and instructed him to participate in training during the coming days. At the same time the trainers are frustrated by the limited effect of three decades of terracing campaigns. The *woreda* officer says:

> *"Now I gave a 15-day-long training course. But the farmers say 'what is new?' They have been told this for 20 years or longer and every time they get rid of the terraces. That's exactly the problem: there is not even one terrace (erkan, Amharic) left to serve as an example. The people in the woreda and the region think it is simple but it is not."*

---

[59] Personal communication Prof. H. Hurni, 21 June 2010. Most households have more than one plot at different locations. Therefore, the floods do not impact all their land.
[60] Personal communication Ato Lakew Desta, 7 November 2011.

*Figure 3.9: Training for terrace construction on the plateau just north of the hill slope in Michael 3 February 2011*

During pegging, when the exact location and direction of a proposed trench is determined while taking account of the slope of the land and the spacing to neighbouring trenches, a second transformation of the terraces takes place. Yonas selected a 55 ha area on which the people of the hill would be working one particular year,[61] knowing that most people do not want the terraces. They not only require a lot of work to construct, but also take up valuable space and make it difficult for oxen to pass during ploughing. Moreover, silt might deposit in the drainage ditches in front of the terraces, and this can lead to their overflowing and flooding the plot. The people of the hill understand that Yonas has to implement the programme because he has been instructed to do so. Several mention that *"it is he who benefits from the terrace, because it gives him good salaries and a* per diem." The space between the terraces is negotiated by the DA, the terrace constructors, and the land users. Although according to the design manual the spacing has to be between 10 metres on steep lands and 21 metres on land with a slope under 8%, a compromise is

---

[61] In the plan in the DA's office, 100 ha are indicated.

reached whereby the space between the bunds differs between 14 metres on steep lands and 34 metres on flat lands (Bhrane 2012). But pegging is also the time at which the DA influences the drainage pattern to reshape relations between neighbouring farmers. By outlining terraces with ditches that cross the boundaries of neighbouring plots, near-horizontal drains are reinserted into the drainage pattern. Every ditch thus connects two or three plots to collective drains down to the river. As Yonas forcefully reinstates collaborations between neighbours who blocked drainage routes and who saw their drainage routes blocked, the terraces come to embody government-prescribed collaboration over drainage.

The digging of terraces creates further compromises in the terracing programme. The first trenches are dug in the presence of Yonas. These are – as indicated in the manual – 0.5 metres deep. As Yonas does not come the third day, both the cross-section of the terraces and the attendance instantly drop. Most terraces constructed are between 0.2 and 0.3 metres deep instead of 0.5 metres. Demaiferam and Tadesse and most other landless people leave after two days. Yohannis continues for a week and is then allowed by the cell leader to stay at home as he is old. The remaining group of landholders gathers for three weeks, but not on Saturdays and Mondays when they too go to the market and not on Sundays and important Orthodox holidays. After 15.5 days of communal terracing on 18 ha, all are instructed to terrace the lands they are cultivating. The cell leader reports every week on progress and participation. None of the *kebelle* officials responds to the low attendance rates.[62] When Yonas passes the lands the following month, he instructs the land users he meets to dig their 'terraces' deeper. All of them promise to do so, but hardly anybody takes action. After the 40-day-long terracing campaign, only a few half terraces have been added outside the initial target area. Yonas reports that 80 ha have been constructed in Yeshat, and this contributes to the *kebelle*'s top ranking on the *woreda's* soil conservation list that year. After that, his boss from the *woreda* instructs him to focus on the distribution of improved seeds.

The first heavy rainstorms of the year expose the season's drainage pattern. In the two months after the onset of the rains, more than 50 people lodge complaints about drainage with the kebelle court and the DA. Some complain about 'terraces' blocked by downstream neighbours. Others complain about damage incurred by new terraces. During light rain showers, the 'terraces' result in more horizontal and thus safer drainage, but

---

[62] Fekadu, one of the farmer leaders, explained: "Cell leaders and *kebelle* cabinet farmers gave permission to their friends and relative to be absent from the terracing work. In return, the absentee farmers thresh grains and plough the land of the cell leaders. Thereafter, other farmers started to complain. They would rather pay the fines than make the terraces they do not want."

when the rainfall intensity is high the water accumulating behind the terraces leads to breakages (Bhrane 2012). These complaints reach the DA. Yonas observes:

> *"People only come to me after the soil conservation season is over and the rains have started. They come to me for everything, even things they can solve themselves. A lot is because their drains are too few or too small. I learnt that many people do not give me complete information. They present their case and hope I will comment. Then they use my words. They say: 'The DA decided this so that is why I am doing it.' Also, if I solve someone's problem, I might create another one. That is why I do not want to interfere. I reject most of the complaints and tell them that if they come next year (before cultivation of the next crop) there is time to construct terraces."*

Although the yearly construction of terraces does not stop erosion in Yeshat, the state terrace construction programme reinstates the 'farmerness' of the Yeshat landholders and the *mengist*'s authority over common property resources management. A yearly cycle has thus been institutionalized in which 'draining terraces' appear between January and March, and slowly disappear again between June and September. The *woreda* government does not question its model of terrace construction. Instead, the persistent erosion is used to justify the need for development committees for soil conservation. Yet, as we saw, persistent soil erosion is not striking a hapless farming community waiting to be saved by modern soil conservation. As the terraces are made and wash away, the hill and the identities of its users take shape.

## 3.7 Conclusion

This chapter documents the morphodynamics of drainage and soil conservation so as to provide a material reading of popular mobilization programmes that make Ethiopia's 'developmental state'. Such a political morphology approach brings into view two elements often absent in political accounts of environmental transformation. First, it recognises how the people of Michael engage in the transformation of the landscape and in the shaping of categories through which this landscape is known by government officials. Of course, they are constrained in their actions: a large group of them does not have access to farm land or oxen for crop cultivation. However, by engaging in lowland trade and establishing social institutions to support this, young men and women in the Choke Mountains perform alternative models to (re)claim their rights. Second, chronicling the changing morphology of drainage and soil conservation networks reveals how the saturation of slopes, the re-routing of drains, and the re-design, layout, and cross-sectioning of terraces redistribute access to land and drainage in ways nobody is able to control.

The analysis of the processes and practices that transform the morphology of a drainage network sheds light on three points about the power and limits of Ethiopia's 'developmental state' model. First, the unsustainable cultivation of the vulnerable slopes of the Choke Mountains is not the inevitable result of a so called 'backward farming community' that is constructed with the model's implementation. Practices of drainage and soil conservation are configured by social and physical boundaries that were established by the distribution of land and the appropriation of rents on the hill. Extraction of crops, taxes and land by subsequent regimes were first accommodated by an expansion of the cultivated area. When this was no longer possible the land was increasingly drained and nutrients in the soil were mined. Involvement of people on the hill in several rounds of - sometimes violent – recruitment for state development programmes and distribution of land undercut relations of sharing land, oxen, labour and drainage. People refused to accept their neighbour's drainage flows, and redirected excess water along vertical plot boundaries, thus increasing its erosive power.

Second, the power of the 'developmental state' model derives from the exploitation of the increasing social and physical tensions on the hill by a new coalition between landholders and government officials. A landholding class of households which received land before the last land distribution in 1997 benefits from a donor- and government-supported agricultural extension apparatus that is geared to uphold the image of 'the farmer in need of assistance for protection of the soil'. Government officials use the ministry's guidelines for community-based participatory watershed development to mobilize influential landowners to organise the construction of terraces in exchange for representative power over 'the community' and support in conflicts over land and drainage. Because the terraces create obstructions that can trigger flooding, landowners convert them into drains. The drainage flows are diverted to plots sharecropped by landless families. Consequently, the yearly mobilization for terrace construction does not halt soil erosion but further aggravates it. Because the landless generation now largely depends on trade or on below subsistence income in lowland areas, they are often not physically present to protect the fields they sharecrop, or to claim their rights. The extension service subsequently labels the young generation,  unable and unwilling to attend its 'farmer tailored' training programmes, as inadequate farmers and holds those absent responsible for the degradation of state land: a criminal act  used to justify forced recruitment for the state soil conservation programme.

Third, the limits of the 'developmental state' model are increasingly visible in the Choke Mountains. While support for a so-called 'community of farmers' creates new solidarities across old political divides, the gullies are expanding into the land of powerful landholders and a generation of young landless families increasingly refuses to contribute labour to a technical development programme that seeks to conserve soils on which it no longer depends. Landless families which fail to live up to the model of the 'farmer interested in soil conservation' have created a competing 'trader model' with its own institutions. The continuous denial of the trader model by landholders and officials fuels generational conflicts over drainage which deepen the fractures in the hill and pose a challenge to government authority. Land degradation thus embodies both the powers and the limits of the 'developmental state'.

# Friction along the canal: Reforming irrigation infrastructure and water user identities in the Gezira scheme in Sudan

## 4.1.    Introduction

*Figure 4.1: Talia (right) and her daughter digging with a hoe to deepen a ditch to get irrigation water, 4 October 2011.*

*Figure 4.2: Gezira main canal at Sennar October 2011. Water is pushed up so high that the water gauge can not be read anymore.*

On 4 October 2011 Ms. Talia and her daughter Eshraga are desperately digging a ditch to get water from the Toman Canal in the Gezira irrigation scheme in Sudan (Figure 4.1). The sorghum and peanuts on the plots they cultivate are wilting as five weeks have passed since the last rain has fallen. Meanwhile, 65 km upstream, more water than ever before is diverted to the scheme from the Blue Nile at Sennar dam (Figure 4.2). So much even, that it exceeds the main canal capacity by 20% and hydraulic excavators are on standby to carry out emergency works in case the main canal would break. With its command area of over 870,000 hectares, 100 years of age and consuming almost a tenth of the entire Nile flow, the Gezira has become an example of inefficient use of Nile water (Karimi et al. 2012). Less than four decades after the World Bank worked with the Sudanese government to double the size of the scheme and intensify its cultivation (IBRD 1966), the same World Bank warns that the Gezira has entered a 'vicious cycle' of poor water delivery, failing water charge collection, poor maintenance, and further declining service delivery (World Bank 2000 p12-13, ibid. 2010). Commenting on the clogged canals of the scheme in a public conference, the President of Sudan labelled the Gezira scheme as "non-feasible" and "a burden on the country's budget" (Dabanga 2014, p1). It seems that the scheme that was long hailed as the engine of the Sudanese economy has broken down.

Since the late 1970s the World Bank and the UN's Food and Agricultural Organisation (FAO) have supported a series of reforms to improve the management of the aging irrigation scheme. In contrast to earlier development cooperation plans which presented both rivers and users as 'in need of development' (IBRD 1959, World Bank 1966), the Irrigation Management Transfer reforms that have been proposed more recently identify these same water users to be in the best position to balance their concerns of food productivity and water security (World Bank 2000, ibid. 2010, FAO 2006). Yet while the government and World Bank focus in the Gezira has shifted to the capabilities and responsibilities of 1,550 newly formed Water User Associations (WUAs) to make the most out of aging infrastructure in distributing a limited amount of water, they have very little to say about the practices through which water efficiencies and productivity are formed. Presenting the farmers as equal water users in a homogeneous yet decaying irrigation scheme provides no entry points for analysing why *Talia* is digging for water on 4 October 2011. Neither can it account for the fact that she does not grow the cash crop that is assumed to be in her best self-interest: cotton. This raises broader questions about who is to be counted as a legitimate or efficient water user and how historical positions of water use come to matter in shaping the physical limits to water supply in the Gezira (cf. Kemerink et al. 2013, cf. Rap and Wester 2017).

To answer these questions this chapter examines how the renegotiation of the friction of an irrigation canal transforms the relative gender, racial and class positions of its users.

Rather than taking categories of water users or guidelines of irrigation management for granted, or highlighting that the linear intervention models that are founded on these categories lack an analysis of underlying social relations of power, I probe both the materialization of recent irrigation reforms and the limits they pose for cultivation in the Gezira. Referring to Anna Tsing's use of 'friction' as a "metaphorical image [that] reminds us that heterogeneous and unequal encounters can lead to new arrangements of culture and power" (2005, p5), I make the friction in the Gezira scheme's canals the subject of our analysis. My particular interest here is not merely in friction as a *metaphor* to highlight the politics of irrigation reform, but also in the *materiality* of the canal as a mediator of changing politico-geographical positions for accessing water along the canal[63]. When sediments deposit or new irrigation routes are excavated, the canal systems changes; at some locations it becomes easier to take water whereas in other places floods or shortages might arise during rainfall or peak irrigation. The canal friction can tell me something about particular patterns through which ideals of water and food security take hold in distributions of water, sediment, crops and labour. And conversely it tells me how these particular distributions shape new ideals of irrigation management. Closely observing the changing irrigation grid, I analyse how new arrangements over the distribution of land, water and labour take form along the canal.

To trace how changing canal friction is formed by and shaping race, gender and class differences, the chapter analyses how Talia's position in the Gezira is dynamically associated with the canal from which she is irrigating. I map how the form of the canal and the relative positions of the users along it co-evolve through changing practices of cultivation and irrigation. Section 4.2 draws from reports by and publications about colonial engineers to analyse how colonial ideas of irrigation development took hold in the canal grid. It shows both how colonial practices of mapping, distribution and intensification were used to concentrate land, water and labour for cotton extraction and how colonial orderings were gradually undermined. Section 4.3 draws on a series of policy documents on irrigation reforms and interviews with people who took part in implementing these reforms to understand how Irrigation Management Transfer (IMT)

---

[63] I am not the first to study how IMT in the name of efficiency and self governance are take their particular form through intersenctions with irrigation 'structures' created by earlier projects of development. For a forceful account of how IMT in Indonesia is shaped by and changed through the Indonesian irrigation agency, see Suhardiman (2013). Other accounts of the politics of irrigation reform that inspired this chapter are those edited by Mollinga and Bolding (2004).

was mobilized to re-establish control over the Gezira's water use(r)s and its irrigation, tenant and worker organisations and thus enabled a new round of extraction of water. Yet as we will see in the remainder of the sections the control over infrastructures and institutions of cultivation is not so easily appropriated. Drawing from measurements of flows of water and sediments, surveys of the shape of the 6 km long Toman canal and interviews with 60 people involved in its use and operation in the 2011 and 2012 summer irrigation seasons, I analyse how the seemingly neutral reforms of free crop choice, privatizing maintenance, participatory operation and sharecropping take hold in the frictions of an irrigation canal. In this way the chapter brings into view how the outcomes of IMT are not singular or prescribed, but the product of ongoing struggles over the distribution of land, water and labour that take form with the changing canal grid.

## 4.2.    Making modern hydraulic properties in the Gezira

The Gezira irrigation scheme was imagined by British engineers in the late 19th century as a gigantic modern enterprise for development of what would become the Sudanese part of its empire (Garstin 1899, Garstin and Dupuis 1904). Soon after the invasion of the Mahdist Empire in 1898 by the British-Egyptian army, plans for the Gezira were further worked out to consolidate Anglo-Egyptian rule. The scheme was not only envisioned as a source of long staple cotton for Britain; practices of colonial mapping, land registration, hydraulic construction, and cotton cultivation were to produce an 'industrious peasantry' that would finance and consolidate control over the Nile territory yet depend for it on a network of British and Egyptian officials and allied investors (Cromer 1908). Yet with the new divisions of land, water and labour taking form in the seemingly evenly distributed grid, existing differences were accentuated and would eventually undermine the organisation of cotton production.

Although the Gezira scheme was designed to incorporate thousands of smaller farms to prevent 'landlordism', prior land holdings and governance structures proved crucial in the consolidation of power by the Gezira gentry. Before the meticulous division of the vast grid of 90 acre rectangles (Figure 4.3), a land survey had allocated holdings to individual families. Plots close to homesteads were allocated to households that used them, and communal lands often ended up in the hands of local leaders, whose support was key for the few hundred British officials who ´indirectly ruled´ over Sudan (Clarkson 2005). After the allocation of land, the government rented it all at a fixed rate, to distribute it to tenants after the construction of the irrigation infrastructure. While an official maximum was set of two 40 acre tenancies per household, people with more than 100 acres were allowed to list "sons, relations, servants or villagers who used or cultivates his land" (Gaitskell 1959, p101).

*Figure 4.3: Detail of design Drawing Canal layout of Wad Numan Block Nr 4 with layout of Toman canal and villages. Ministry of Irrigation Undated (every rectangle represents 90 acres of land – the total current command area of the scheme is 2.1 million acres). (Source: Ministry of Water Resources, Irrigation and Electricity, Wad Medani Gezira map archive)*

The regular distribution of water over the Gezira plain proved equally illusionary. The British engineers used Manning's (1895) experimental formula which captured the relations between the form and slope of clay canals and the water discharge to design cross sections of the canals that would distribute the water over thousands of equally sized farms (Dooge 1992). The application of the formulas was far from straightforward for two key reasons: 1) silt in the Nile and 2) irregularities in the slope of the land. Unlike the experimental channels used by Manning to derive his formula, the discharge in the canals changes over time due to sediment deposition. Moreover, higher areas made the regular distribution of water through straight irrigation canals impossible. From the outset struggles erupted over canal maintenance and the need for *nacoosis* – additional intakes that did not fit the irrigation grid – to get more water to areas that were difficult to irrigate. *Ghaffirs* – canal operators – would use regulators in the minor canals to regulate the water level over the canal (Ertsen 2016). The continuous work to get the water to the field outlet pipes would wield considerable power to the bureaucracy of engineers and *ghaffirs* (operators) of the Ministry of Irrigation that was set up with the scheme in 1924[64].

Perhaps the biggest challenge to the colonial Gezira project was not the control of land or water, but to find labour for construction and cultivation. The central idea of the scheme was that tenants and their families would cultivate their own smallholder tenancy. To create "a sense of responsibility & cooperation in the native" and a "unity of interest between the Government & Syndicate", wrote the manager of the Scheme in 1959 (cited by Ertsen 2016, p92), every land user, through his tenancy, was to become a shareholder in the venture of cotton production. The modern property relations between the tenants, government and the private syndicate for cotton cultivation were formalized in the 'profit sharing agreement', which states that profits from cotton sales of a tenancy would be divided in the following way:

> "35 per cent to the Government. To cover interest on the loan, amortization, and maintenance of irrigation works and canals, and rent to the natives for the lease of their land. 25 per cent to the Syndicate.[65] To cover the cost of roads, drainage, subsidiary canals, clearing and levelling, agricultural supervisory staff, accounting staff, and Syndicate's profits. 40 percent to the tenant. To cover the cost of labour, seed, agricultural implements, use of tillage animals, and tenant's profit." (Kitchener 1913 in Gaitskell 1959, p70).

Most tenant families did not have sufficient labour for the intensive cultivation of cotton. To solve this problem the Anglo-Egyptian government worked hard to attract people displaced by other projects of British colonialism in the Fulani Sultanates – and from Chad and Western Sudan (Duffield 1983, O'Brien 1984, Clarkson 2005). Apart from providing transport to the scheme and separate areas for them to settle within the grid, the strategy hinged on a combination of creating needs for cash through taxation and the introduction of consumer goods while repressing the development of wage labour markets elsewhere in the country (O'Brien 1983, p19). By 1928 80,000 migrant 'labourers' had settled in the Gezira (O'Brien 1984a, p124).

The irrigation grid shaped labour relations of the Gezira along the geometric, racial and gender lines. Male 'Arab' tenants organized distribution of water (irrigation) and labour.

---

[64] For a fascinating history of how irrigation engineers, Syndicate inspectors, operators and farmers negotiated the distribution of water and cotton in the Gezira in colonial times, see Ertsen (2016).

[65] The role of the private 'Sudan Plantations Syndicate' was taken over by a parastatal called the Sudan Gezira Board in 1950. From 2010 it was renamed the Gezira Scheme.

They hired 'migrant'[66] men, women and children for the construction of soil bunds, weeding, and the application of fertilizers against cash payments. When the cotton crop was not attended to the liking of the Syndicate inspectors, the Syndicate itself would hire migrant labourers to do the work and charge the work for higher prices against the tenant's share of the profit (O'Brien 1984, Abdelkarim 1992). Migrants thus became both the tenant's main source of debt and a tool of tenant coercion.

Tenant debt for hiring labour was at the root of another widening split that would be decisive to the cultivation of the Gezira. While foreign investors and rich tenants benefited from investing in cotton cultivation, a large part of the tenantry went into structural debt with their peers to finance operations (Barnett 1977). As early as 1937, a journalist described how a tenant struggled to pay the wage bill: "A wave of doubt submerges him when he sees that the Syndicate's 20% share results in palaces and luxury for the Syndicate, while he suffers from the greatest penury in spite of his 40 per cent" (Al Nil 1937, quoted in Barnett 1977, p121). The increasing number of foreclosed tenancies which were handed over to migrants only added to tensions within the tenantry. To contain tenant unrest and to ensure continued tenant loyalty (and cotton supplies) after independence, the Scheme Management and the government pushed through a number of reforms in favour of the Arab tenant population (O'Brien 1984). The Sudanization law (1948) forbade land holding, salaried jobs and school attendance to non-Sudanese (Gaitskell 1959, p313), a category which was narrowly defined as everybody whose ancestors did not live in Sudan before the British occupation in 1898 (Duffield 1983, p57) and was conveniently applied to all people with a dark complexion, including those from Darfur (West Sudan). By 1952, a Tenant Fund was set up to provide facilities like wells and schools, which would later enable their offspring to obtain salaried jobs in towns. The Gezira Board, which took over the role of the Syndicate in 1950, made very clear that its services were to benefit the tenantry in its 'rule number one' for new staff members of the scheme: "The tenant is a tenant and not a labourer" (SGB 1951). These discriminative policies could not prevent increasing poverty and divisions among the tenant population. In 1946, a mass strike of tenants demonstrated for the first time the potential of tenant organisation, when 75% of the tenants refused to plant the cotton crop for 6 weeks and demanded and obtained a larger share of the cotton profit (El-Amin 2002, p164). Not satisfied with their representation through the Tenant Association which was set up by the scheme management, tenants set up an alternative Tenant Union with the support of

---

[66] While generally indicated by tenants as Fellata, I avoid this term as it is considered derogatory by the people of Shazera. I use the term migrants to indicate both those who moved to the scheme from Nigeria, Chad, and West Sudan and their descendants.

the communist party to further push their demands (Clarkson 2005)[67]. In this way industrial cotton production in Gezira scheme not only institutionalized a highly divided irrigation society of tenants and workers but also concentrated a large impoverished tenant population which would grow into what some argued the most powerful union in Africa (First 1970).

Despite mass uprises, the Gezira model continued to be promoted as a model for development of African land, water and people after independence (1956). In the late 1950s and early 1960s the World Bank provided loans for the Roseires dam and the Managil extension to almost double the Gezira scheme's irrigable area (IBRD 1963). Moreover, the completion of the dam and the introduction of pesticides and aerial spraying of herbicides enabled an increase in the cropping intensity from 50% to 75% (Wallach 1988). When world cotton prices dropped in the late 1950s, both the government and tenants struggled to finance operations. Decreasing profits and lack of investments in cotton cultivation made tenants shift to cultivating crops for their own consumption and local markets. Confronted with reduced income from cotton proceeds and powerful infrastructures, bureaucracies and labour organisation that were produced with the earlier divisions of land, water and labour for cotton cultivation, the World Bank, IMF and the Sudanese government sought to rearrange modern hydraulic property arrangements of the Gezira by Irrigation Management Transfer.

## 4.3.    Irrigation   management   transfer  –  rearranging   modern hydraulic property

The gradual undermining of the profit sharing agreement led to an apparent reversal of the colonial view of water users as 'backward', and later, 'in need of development.' From the early 1980s, irrigation reforms were redirected towards reducing state control and public debt by enhancing entrepreneurial ingenuity of the Gezira tenants. The World Bank proposed the transfer of irrigation management to water users as a logical solution to realign modern hydraulic property relations to transforming networks of domestic and global trade and power (1982). Over the decades that followed a programme of Irrigation Management Transfer (IMT) was unfolded which not only served to dismantle the colonial irrigation and labour bureaucracies but also would renew the terms for

---

[67] The Tenant Union had a crucial role in bringing down governments in 1956, 1964 and 1985 (Ali 1989, p117, Niblock 1987).

marginalization of cash-ridden tenants and workers or free up water for 'more productive' uses.

The World Bank argued that the global 'underperformance' of the large irrigation schemes was the result of ballooning irrigation bureaucracies (World Bank 1982, World Bank 2000, World Bank 2010). To prevent defaulting of loans and losing control over government, new loans were conditioned on the liberalisation of production of tropical exports like sugar and cotton. To increase incentives for farmers to invest in the production of export crops, constraints posed by government agencies, irrigation bureaucracies and unions on the organisation of land, water and labour were to be removed. The idea was first, that the continuation of production for export would increase economic growth for southern countries and enable them to repay their loans (IMF 2014); second, it would ensure a growing market for machinery and fertilizers and for cheap wheat produced on farms in Europe, the USA and Australia (McMichael 2012).

The liberalization of irrigation was justified by a surge in theorizing on property rights. Economists highlighted the high transaction costs of the (post) colonial contracts, designs and bureaucracies on which developmentalism was founded. Their new liberal ideas identified the commitment of users, not the regulation by state, as the most important basis for efficient allocation of resources. When left to users themselves, Coase and Demsetz argued, property rights will emerge whenever these will reduce social costs (Coase 1960, Demsetz 1967). Social scientists involved in agricultural development drew from their 'new institutional theory' to analyse the creation of property management and rights in community managed irrigation schemes in the Asia and Africa (Coward 1986, Ostrom 1992). Instead of seeing water users as mere implementers of roles assigned by the bureaucracy, they regarded user involvement in construction, operation and maintenance as central to the creation of property relations. On the basis of 153 studies of Asian irrigation schemes, Coward conceptualized irrigation as a hydraulic "property creating process [with] two linked meanings: (1) the creation of new objects of property and (2) the possibility of new property relations" (Coward 1986, p492) to derive "policy variables to be used in designing further irrigation development activities" (Coward 1986, p494). In a similar fashion, Ostrom drew from new institutional theory in her analysis of 135 case studies in Nepal to theorize design principles for sustainable irrigation management organisations (Ostrom 1992).

The empirical methods they used for analysis of the performance of irrigation resembled the scientific method used by the engineers to derive empirical formulas that were used to design irrigation canals of the Gezira a century earlier. Like Manning's formula for open channel flow, the design principles and policy variables derived by Ostrom and

Coward were derived from numerous empirical observations. And like Manning had stressed that his formula could not be used outside the material domain in which it had been established (Manning 1895, cited by Liu 2014, p136), Coward stressed the specificity of the hydraulic property creation process (Coward 1986). In the end , modern empirical sciences were all about abstraction to 'empirical formulae' (Manning 1895), 'design principles' (Ostrom 1990) and 'policy variables' (Coward 1986, p494). Where Manning's formula assumes uniform flow and says nothing about the dynamics of sedimentation, the principles of clearly defined boundaries, graduated sanctions and proportional investment in maintenance identified by Ostrom and Coward provide no entry points for analysing the differential constraints of accessing land, water and labour that shaped crop production and infrastructure. Likewise, FAO's IMT principles construct 'water users' as equally positioned to improve the distribution of water. Meanwhile the implications of the particular use of these principles on forming and perpetuating the differential positions of water users in the Gezira have been left unanalysed.

The first irrigation reform in the Gezira was to replace the colonial 'profit sharing agreement' with the 'individual account' system. The World Bank provided a 60 Million US$ loan package of machinery, fertilizers and pesticides on condition that the government would end its role as shareholder in the venture of cotton cultivation (World Bank 1979, p39). The central idea was that instead of receiving a share of the profits of cotton production, the government would be compensated for the land and water services it provided by a system of flat taxes. After deduction of taxes and services from this 'individual account', the tenant was to receive the full profit of cotton cultivation. The equation of tenants with water users implicit in the individual account system – like the colonial profit sharing agreement – made invisible the role of more than a million of 'migrant' workers who took over most of the work in the scheme, thereby contributing to their further marginalization. By positing that entrepreneurial risk would lead to increased production by natural selection of the 'best' tenants, it assumed that tenants would share a common goal of profit maximization, and failed to appreciate how tenant 'performance' was shaped by dreaded systems of land preparation, money lending, cotton marketing and currency conversions over which they had little influence[68].

Over the decades that followed the reforms progressed and took the form of a fixed recipe of Irrigation Management Transfer that was implemented in more than 60 countries (FAO

---

[68] This is why the Tenant Union called for a strike to plant cotton in 1979. After heavy lobbying with the union leaders and small concessions to payments to the tenants from the tenant fund, the Tenant Union gave in the year after (Ali 1986).

2007). In its guidelines of more than 100 pages, the FAO summarized IMT's three main principles as follows:

i.   the water users' association should be in the driver's seat (identifying, prioritizing and making the financial decisions);

ii.  the irrigation agency should facilitate and provide technical assistance, not direct the process;

iii. future infrastructure improvement should exemplify a farmer-driven, incremental approach rather than the typical fully subsidized, non-participatory approach of the past. (FAO 1999, p90)

In the 1990s – when Sudan defaulted on some of the payments and land conditions, the IMF and World Bank blocked further cooperation – president Al-Beshir had already started working on the second point of the agenda by downsizing the irrigation and agricultural bureaucracies that had undermined power of his predecessors. Agricultural services like land preparation and fertilizer provision were increasingly privatised. The number of people working for the scheme was reduced from over 13,000 permanent posts and around 25,000 seasonal posts in the 1970s to 7860 posts in 1998[69]. When the Sudanese government resumed cooperation with the World Bank in 2000, the first joint activity they engaged in was the completion of the IMT agenda for the Gezira. In its Options for Development report the World Bank wrote:

> "While the original design of the Gezira Scheme envisaged centralized management, it is possible to move to a decentralized management system in which production decisions and the provision of services are left to farmers and the private sector. This shift would not require special modifications to the existing physical infrastructure" (World Bank 2000, p iv).

Within a year after publishing the report a pilot was set up with support of the World Bank and FAO which introduced Water User Associations in 5,000 ha of the scheme to take control over operation and maintenance activities. Soon after the pilot was proclaimed as successful (Abdelhadi et al 2004, FAO IMT country profile cited by Vermillion 2006) the president signed the Gezira 2005 act, which instructed upscaling of the pilot to the rest of the scheme. In 2009, the railways, ginneries and agricultural engineering department were sold. In January 2010, the government ordered the transfer of operation and maintenance of the minor canal to 1,550 Water User Associations. In the same year, all remaining employees of the scheme were fired, and replaced by 75 people

---

[69] Interview HRM manager Gezira Scheme 6 December 2012.

in charge of the Agricultural Research station, Extension and Training and supervision of Finance and Administration of canal operation and maintenance.

Assuming that tenant control over land and water would ensure an efficient allocation of resources, IMT in the Gezira carefully scripted who are legitimate users and what are appropriate structures for efficient water management. The more than million migrant workers like Talia do not feature in the Gezira act, the World Banks options for development or the success story of Al Hakm pilot with Water User Association in the Gezira. The persistent focus on the 'official' 120,000 tenants of the scheme, now formally in control over which crops are grown, who operates the canals, and how much is invested in maintenance, undermined not only the bureaucracy with whom they engaged in the cultivation of cotton but also further marginalized the labourers to whom the tenants transferred most of the cultivation.

Yet the relations binding together 2.5 million people to a $10^{th}$ of the Nile water were not so easily removed. Like the colonial profit sharing agreement, the implementation of IMT would thus produce specific limits to the ways in which flows of water and silt are organised today. In the following sections I turn to the Toman canal to get a grip on how these particular limits are formed and what this means to whom along the canal.

## 4.4.    Cropping patterns – the limits to free crop choice

What struck me at first when I arrived at Toman in 2011 was that less than 10 % of the irrigable area was cultivated by cotton (Figure 4.4). Most of the area was grown with low value sorghum crop, which seemed to be almost exclusively cultivated by migrant women like Talia. Was this the result of the liberalization of the cropping pattern that would bring more high value crops?

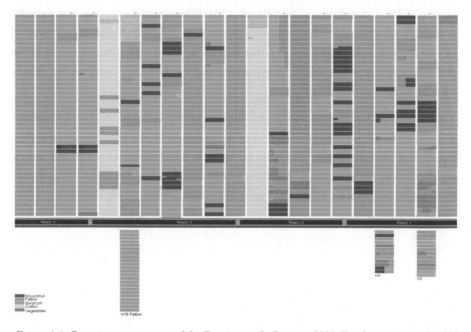

*Figure 4.4: Cropping pattern around the Toman canal - Summer 2011. Total area represents 800 ha (Brown = fallow, Green = sorghum, Purple = groundnuts, Yellow = cotton, Red = vegetables, Grey = village). The Canal is indicated in blue and flows from the East to the West.*

To answer this question I turn to the tenancy of Mr. Mahmoud to analyse how access to land, water and services take form over the Summer 2011 irrigation season. Mr. Mahmoud is the tenant whose plot is cultivated by Ms. Talia. Another tenant, Mr. Elfatih, provides credit for cultivation and controls machines that work on most of the Toman lands. Mahmoud's tenancy *(hawasha)* consists of 5 plots of 4 *acre* (1.68 ha) each. The plots are situated at the middle of the 10th, 11th, 12th, 13th, and 15th field canals tapping from Toman canal (Figure 4.5). Each field canal supplies 1 *nimra* of 90 *acre* (38 ha) land.

Each *nimra* is divided in 2 *acre* and 4 *acre* plots depending on whether it was split up one or two times since the construction of the scheme.[70]

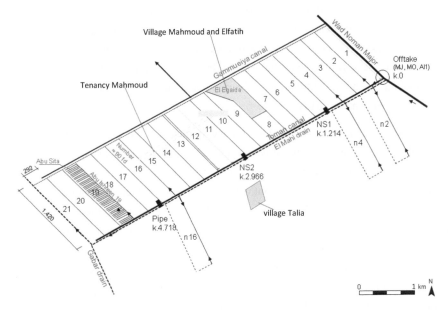

*Figure 4.5: Toman canal and the 21 nimras irrigating from it. South of the canal are three nimras which are also irrigating from Toman canal . Figure by Emma Aalbers*

## Mahmoud – a tenant/debtor

For Mahmoud (age 58) the cultivation of the tenancy was never a lucrative business. The credits he needs to finance the cultivation of cotton and wheat make him reluctant to invest. Until 2006 the Gezira board determined the cropping pattern and at least advanced periodic credits over the season to pay for the cultivation. He recollects how:

> "Before, the Sudan Gezira Board provided all credit, machines, fertilizers, and spraying. I cultivated and they did the services and administration. They were

---

[70] In 1924 each *hawasha* with 3 plots of 10 *acre* had been handed out to individual tenants. In 1931 this was changed to 4 x 10 acre (Gaitskell 1959). Over the century the rotation was changed several times to eliminate disease and incorporate new crops before the five course rotation was introduced to incorporate livestock into the scheme in 1991 (see Gaitskell 1959, GoS 1992 (Irish project to integrate livestock)). With the change of the rotation and the splitting of tenancies upon inheritance the tenancy size changed. In Gezira the common sizes for a tenancy are now 5 x 4 acre and 5 x 2 acre.

strict: they adjusted all the advances in the cotton payments. When there was not enough yield, they transferred the debt to next year. When a lot of debt accumulated sometimes they would waive it however. But there are no payments for labour during the season anymore. We have to finance the labour ourselves and we need to settle the bill straight away. If you can't pay the bank after the season they send the *eshtekik* [scheme police] after you. They are looking for me ... Cultivation is for rich people now" (Interview Mahmoud Ali 23 November 2012).

Over the years, cultivation has become a secondary income to the tenants of Toman. Most children of Toman tenants left cultivation altogether. Many of them, Mahmoud's sons included, attended post-elementary education in nearby towns or Khartoum. Often the tenancy is left to one child of the tenant who remains in the village. Mahmoud's sons have both become taxi drivers in Khartoum. One of them was involved in the management of the farm while living 250 km away in Khartoum. A few years ago Mahmoud's wife started a small shop in their house. From 2000 Mahmoud found a job selling bus tickets in the nearby town of Wad Medani. The cash income generated by these activities solved part of Mahmoud's cash flow problem. But he uses two other strategies to finance cultivation of his tenancy: He goes into debt with Elfatih and he sharecrops out his sorghum and peanut plots to Talia.

## Elfatih – a tenant/ money lender

Elfatih is the richest tenant of the village and his tractors plow Mahmoud's land on credit without interest. His grandfather was a local leader who owned rainfed land around Toman before the scheme was built. When the Toman canal was dug in 1924, the (Anglo-Egyptian) government rented his land and gave him three tenancies. The family owned a large number of cows and managed to increase its wealth. When state investments declined and services faltered in the late 1980s, Elfatih bought tractors and combine harvesters. While he and his family still earn well from their tenancies and the plots they rent in from others for cultivation of vegetables (year round), sorghum (in summer) and wheat (winter), they make most of their money from land preparation, credit provision and a gas station they opened up in the nearby town.

Steadily expanding his arsenal of machinery, Elfatih was asked by members of the ruling party to become the leader of the local branch of the new Farmer Union and facilitate negotiations between farmers and service providers. When the committee for the Water User Association (WUA) had to be selected for Toman canal in 2009 it was only logical that he picked the candidates from his 'home' area and almost all other tenants agreed with his proposal. Elfatih made the *wakeel*[71] of his wheat and sorghum crops the WUA president and became treasurer himself.

When cotton prices went up in 2010 and the Sudan Cotton Company[72] (SCC) wanted to grow cotton in the Toman area in 2011, the company approached Elfatih. He arranged a WUA meeting to inform all tenants that they had to participate in the cultivation of cotton as it could only be effectively grown by the whole *nimra*. The farmers were paid 75 SDG (US$ 12) and one sack of fertilizer per acre for the cultivation of food crops on other plots upfront and were promised 800 SDG (USD$ 128) per *kantar*[73] cotton harvested. Because the prospect offered was good, 27 out of 30 tenants who had land in the *nimra* participated. Without further discussion Elfatih's tractor thereafter ploughed *nimra* 10 of Mahmoud's tenancy.

## Talia - a migrant/sharecropper on Mahmoud's land

Because Mahmoud and his family members all work outside the farm and they go in debt to pay for cultivation of the cotton plot, they have no labour and money to invest in other plots. Like most other tenants Mahmoud therefore sharecrops out two of his five plots to Talia for growing sorghum (6 acres) and groundnuts (2 acres). Three years ago Talia, her husband and four children moved from Darfur to Shazera which is just across the Toman canal. Shazera was established as a migrant village when the scheme was constructed.

The canal separates the immigrant village from the 'Arabs' village in which Elfatih and Mahmoud are living in brick houses tenants. Sharecroppers live in mud houses. Electricity, tap water and schools have not reached Shazera until today. The engagement of people in Shazera in irrigation differs by gender and generation. It is not a coincidence that we encounter a migrant woman and her daughter digging with a hoe in October. Outside the harvest season the work on most fields around Toman is done by migrant women and children. While most men in Shazera spend part of the year outside of the Gezira for wages, the work on the fields around the village reinforces the women's 'logical' roles of taking care of the homestead and children. While Talia cultivates the

---

[71] Manager who worked for him: he arranges labour, inputs, administration etc.

[72] The SCC had been founded when services, management, spraying collection and ginning were privatised in the 1990s. Its shareholders are 'the tenants of Gezira, Rahad and New Haifa' who are organised by the Farmer Union, the Farmers' Commercial bank and the National Pension Fund (Sudan Cotton Company 2014)

[73] 1 kantar = 157 kg seed cotton;

land with her 15, 13 and 7 year old daughters who have never attended school,[74] her husband is working on sesame farms some 200 km away in Blue Nile State.

Only few men between 20 and 50 from Shazera stay in the village for most of the year. While three of them have a tenancy the rest of them work on the land of Elfatih and two other rich tenants living in the Arab village. They work in vegetable fields mostly for wages. The sorghum crop is often cultivated by people of Shazera on a sharecropping basis. While at first sight the arrangement for sharecropping by Shazera men looks similar to Talia's - like her, they sharecrop on a 50-50 basis, i.e. the yield is split between the provider of the land and the labourer - in practice the terms are very different. While Mahmoud does not involve in cultivation at all and Talia grows without inputs or supervision, the men employed by Elfatih are supervised by *wakeels* (caretakers employed by Elfatih) who check if the crops are sufficiently attended to. Not only does their close field supervision make this type of sharecropping inappropriate for women; the higher degree of mechanization and attendance to the crop make it similar to a wage labor job transforming it into a man's job. One of the men working for Elfatih explains that even though after deduction of machinery, management costs and inputs, the final share obtained at the end amounts to perhaps only a fifth of the yield, the combination of high yield and the relative reliability make this a preferable venture compared to the sharecropping done by poorer tenants.

Of the 150 cultivated plots of the rotation[75] we closely observed in the summer of 2011, 40% is sharecropped to migrants who grow sorghum and groundnuts with a limited amount of inputs. The high intensity model of supervised sharecropping we only encounter in the neighbouring rotation, where Elfatih has his land. On the remainder of the sorghum plots and most of the plots with cotton and vegetables, the management of cultivation is in the hand of tenants themselves or their caretakers (45%) or other tenants who rented the plots from them (5-15%). On many of them migrants work for wages on activities like fertilizer application and weeding.

The irrigation reforms have thus not, as envisaged by the World Bank (2000, p52), led to the transfer of land from tenants trapped in debt to profit making tenants. Neither has this led to a seamless monoculture of mechanized cotton cultivation. Instead there is a lot of sorghum and a high variety in the arrangements of cultivation. Indeed the liberalization has enabled further appropriation of cheap labour by rich tenants who use sharecropping

---

[74] She is thinking of sending her 5 year old son to school.

[75] The rotation I closely observed consists of 30 tenancies, each consisting of 5 plots along 5 field canals (numbers 10, 11, 12, 13 and 15).

as a way to increase the incentive for those who cultivate the land and draw on their family networks at times of high labour demand. A large share of the poorer tenants, who have no money to invest in agriculture, have resorted to a 'low investment' model of sharecropping, largely by migrant women and children. They used their 'privileged' tenant position formed by education in tenant schools to become wage labourers in the city but did not sell their tenancies. Sharecropping enabled them to hold on to the tenant identity that was carefully cultivated with the grid, without investing in labour or engaging in hard field labour themselves, which they consider inferior. The transfer of control over land and labour to those who do not enjoy the hard-won tenant rights of representation in the Gezira board, education and access to safe drinking water not only goes against the economic rationality assumed in the options for development report; it undermined the colonial class basis which distributed managers and workers of the land on different sides of the canal. This helps to explain the prevalence of sorghum for which limited capital investment is required and of which crop yields remain so low that Elfatih refuses to cultivate it. While crop choice in the Gezira is officially 'free' for the tenants since the implementation of the 2005 Gezira Act, the majority of tenants hardly feel this as such. For Mahmoud the cotton grown in *nimra* 10 is not his choice, but the choice of Elfatih and the Water User Association. And while Mahmoud technically decides that *Nimra* 12 and 15 are cultivated with sorghum and groundnuts, he had really not a lot of choice in picking these crops either. With limited money and labour to invest, he sharecropped to Talia who grows the crops without additional inputs[76].

The (re)organisation of labour provides a start, but does not suffice, to explain the choice for sorghum by sharecroppers like Talia. In the next sections I turn to the changing canal form to explore how the limits to the cropping pattern emerge through, and produce new, social and physical boundaries of irrigation.

## 4.5.    Shaping Toman canal

While we saw that the cropping pattern is no longer strictly prescribed by irrigation institutions set up in colonial times, the old canals seem to have remained firmly in place. Yet when taking a closer look, we find that Toman canal too has profoundly changed. The most obvious traces thereof are found along the first reach the canal. Here the embankment has grown 10 meters wide and 2.5 meter high. The canal bottom in the head

---

[76] Mahmoud sold the fertilizer for cash. His plots in the other two *nimra*s are fallowed.

end of Toman canal is now a meter higher than indicated in the 1924 design (Figure 4.6). Further along the canal the size of the embankment steadily decreases (Figure 4.7).

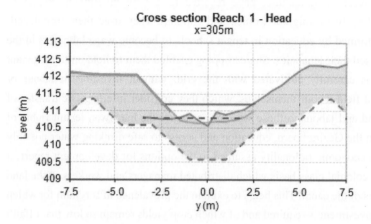

*Figure 4.6: Cross section of Toman canal head end surveyed by author in September 2011. The red dotted line indicates the profile of the original design. The shaded area is the deposited sediment. The blue dotted line indicates the original full supply water level of the canal. The continuous blue line indicates the average water level in October 2011*

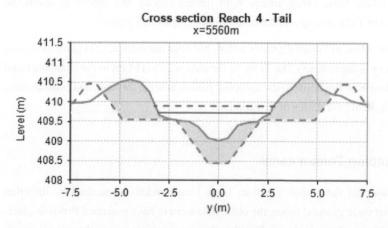

*Figure 4.7: Cross section of Toman canal tail end September 2011. The red dotted line indicates the profile of the original design. The shaded area is the deposited sediment. The bue dotted line indicates the original full supply water level of the canal. The continuous blue line indicates the average water level in October 2011*

The dark brown colour of the water in July reveals the origins of the rising canal bottom and banks. In upstream Ethiopia the summer rains erode the freshly plowed lands, so that the Blue Nile carries a heavy load of silt. As a result, some 8 million tons of sediment enter the Gezira every year. While 40% ends up with the water in the irrigated fields, almost a similar amount is trapped in the minor canals (Gismalla 2010). When water enters the minor canal the velocity drops[77] and part of the sediment deposits. The 10 meter wide embankment alongside the 5 meter wide canal is the product of a century of canal maintenance. Over the lifetime of the scheme the intake of sediments has accelerated for two reasons. First, with the diversification and intensification of the cropping pattern in the 1960s and 1970s the water consumption had increased by some 50%. To accommodate the rising cropping intensity, the start of the irrigation season was brought forward by two weeks to the end of June. Second, increasingly intensive cultivation in the upper Blue Nile has led to a fourfold increase of silt content of the water entering Gezira (Figures 4.8 and 4.9). Precisely when people start to take large amounts of water for planting of the crops between 5 and 25 July, sediment concentrations of the water entering Toman canal peak (Figure 4.10). An estimated 4,000 ton sediments entered Toman canal between the start of the irrigation season on 27 June and the moment the intake of water was halted because of heavy rainfall on 4 August 2011.

---

[77] Because the minor canals were designed as storage canals in which the water velocity is lower than in the major canals which are designed only for conveyance of the water.

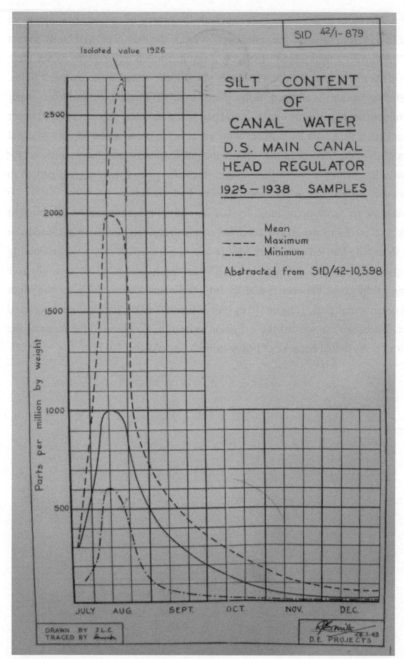

*Figure 4.8: Silt content of water in the Gezira main canal 1925-1938 Source: SID 1943 (1000 ppm by weight means that with every litre water 1 gram of sediment is entering)*

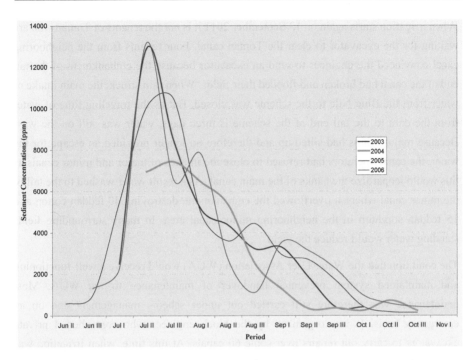

*Figure 4.9: Silt content of water in the Gezira main canal 2003-2006 Source: HRS 2008 [Note the difference in scales of the Y-axis in Figures 4.8 and 4.9]*

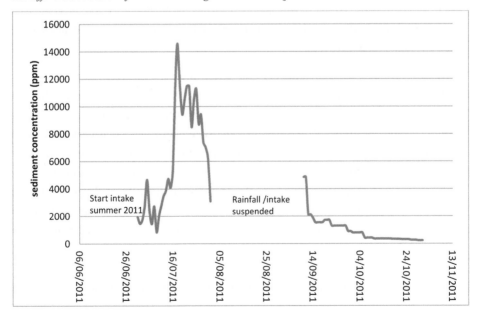

*Figure 4.10: Silt content of the water entering Toman canal Summer 2011 - prepared from daily samples at Toman offtake collected by Eshraga Sokrab.*

When irrigation starts again on 13 September 2011 it is *not* the tenants of Toman who are waiting for the excavator to clear the Toman canal. Four tenants from the neighboring canal convinced the engineer to send an excavator because the embankment of the tail end of the canal had broken and flooded their fields. When rain struck, the main intake of water from the Blue Nile to the scheme was closed. But as the traveling time of water from the dam to the tail end of the scheme is three days, water was still on the way. Because major drains had silted up and therefore no longer provided an escape for the water, the canal operators had refused to close the intakes of major and minor canals as this would jeopardize the banks of the main canal. As a result water washed to the tail of the minor canal where it overflowed the embankment, destroying 40 feddan cotton and 15 feddan sorghum in the neighboring minor canal area. In many surrounding fields standing water would reduce the yield.

The condition that the Water User Association (WUA) would receive a well-functioning and maintained system prevented handover of maintenance to the WUA. Most maintenance was therefore still carried out under scheme management and on an emergency basis. It was organized by the block engineer who controlled 4 private excavators to carry out repairs over some 60 canals. At this time, when irrigation was started up again after the rain and almost all plots wanted water, his office was swamped with people demanding an excavator. The engineer explained that 3 out of 4 excavators were broken because of spare parts and maintenance problems. Therefore he contracted an excavator from Ali, a tenant in the area north of Toman, who had just purchased three new hydraulic canal excavators.

The entry of Ali's company was only the latest in a series of reforms to get private companies involved in canal maintenance. From the 1990s the government had pulled out of maintenance when it transferred the financing of farm operations to a number of private banks. A lobby of influential businessmen and state politicians had persuaded the vice president to 'resolve the problem of cost recovery' through reducing costs by allowing private companies to apply for contracts for desilting of canals (1997) and transferring the maintenance of minor canals from the Ministry of Irrigation to the Sudan Gezira Board (1999). In the same year a group of influential tenants initiated the creation of a large new private excavation company which was granted a contract. Its shareholders were the Gezira State, the Sudan Cotton Company and a large group of private investors[78]. They bought 40 new excavators from Korea with a credit from the Gulf and got a license for excavation of a third of the scheme. In its first year the company cleared 42 million

---

[78] Gezira state 19%, Sudan Cotton Company 20% and Asalma 61% private investors

cubic meter of silt from the canals and paid a 300% dividend to its shareholders (personal communication former director of the company, 4 December 2012 ) (Figure 4.11).

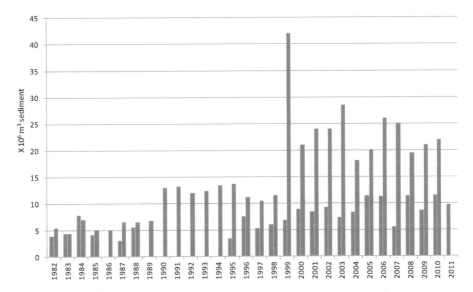

*Figure 4.11: Plot of records of amount of sediment that entered the scheme (blue) and were excavated from the canals (red). Curiously since 1995 the reported amount of sediments removed from the canals is much larger than the amount reported entering the scheme. Data sources: sediment entering the scheme (HRC 2011 - no data for 1989 to 1994); excavation record: Abdul Salam (2009) and The Gezira Scheme (2012)*

Sediments depositing in the canals were thus turned from a burden into a lucrative business. While this business is now for 90% in the hands of private companies, the water charges introduced after abandoning the profit sharing agreement to cover for operation, maintenance and overhead have never sufficed to pay the bill for canal maintenance and the state has continued to cover the shortfall. By 2011, 44 companies were involved in the excavation of Gezira canals. The previous year Ali had managed to buy a package of 3 excavators, 4 tractors, 3 trucks and a grader worth 2.4 million USD. The machines were financed by the Agricultural Subsidy fund of the Ministry of Agriculture. Ali paid 5% of the machines up front through selling old machines and got the other 95% on credit. It turned out to be a good investment as the scheme guaranteed him work for his machines.

After repairing the breach, for which the operator of the excavator was paid by tenants from the neighbouring canal area, the excavator moved up to desilt part of Toman canal. As the excavator is paid by the Gezira Scheme for the canal length excavated, the operator did not seem too concerned with the shape of the canal. Without surveying the level of the bed, he scraped some 30 cm from the bottom to strengthen the embankments of the tail end and improve the flow before rushing to the next canal.

The silted cross-section of Toman canal then, is not just the 'natural' result of increasing erosion in Ethiopia or an inefficient operation and maintenance bureaucracy. The physical boundaries of the canal emerged through the deliberate reorganisation intensification and privatisation of maintenance. As more and more water is pushed through the increasingly silted canal network, and excavators are no longer concerned with the shape of the canal but with the volume they register for excavation, irrigation in Toman is profoundly reorganised. In the next section I examine how the new canal form comes to matter in the reorganisation of irrigation.

## 4.6.    Distributing water – making a difference in water access

After irrigation from Toman canal is restarted, it soon becomes clear that the users are not equally positioned in accessing water. While farmers along the first reach of the canal can access water right away, for Talia it takes 24 more days until she can irrigate. Who is in a position to access water has much to do with the thick layer of sediment on the canal bottom. Because the original cross section of the minor canal had been largely filled with sediment the water storage capacity of the canal is strongly reduced. This made that sudden openings and closings of field canals by the users result in large fluctuation of the water level. As control of water levels in the canal had thus become extremely difficult, the *ghaffir* (gate operator) who was until recently in charge of canal operation had fixed the regulators so that the canal would gradually fill from the head to the tail end. In this way he only had to check if the water level in the last reach was too high or too low and tell his colleague of the irrigation department at the beginning of the canal to adjust the flow accordingly. When the irrigation department had been closed in 2010 and 2,000 ghaffirs went on strike the routine visit to check the tail water level ended. As water control in the minor canal was now assumed in the hands of the users, the scheme instead hired 1,700 new agents from a private security company, but mainly for tax collection.

Now that everybody had finished weeding after the rain storm everybody needed water and all 16 Field Outlet Pipes of cultivated *nimra*s are open.[79] The water intake into the

---

[79] A similar situation occurs during planting. These might sound like exceptional periods. But for Talia's sorghum plot, and many other sorghum plot these are the only two (and therefore crucial) irrigation turns.

scheme had been fully opened and in the major canal the level was operated upto the banks to push it through. After entering the minor canal the silt pushes the water up almost a meter above the design water level. As a few centimetres already make a big difference in gravity irrigation, the water bursts into the first field canals. Because a lot of water is taken out in the first reach of the canal the water level quickly drops along the canal. The trickle that passes through the gate opening is insufficient to fill the next reach and flows to the tail end of the canal where the water level starts to rise slowly.

This is why along the 65 km of canals from Sennar dam to the first regulator in Toman all structures were overflowing, while two kilometres further down, at Nimra 15, the investor who rented the first few plots along the minor canal had brought a diesel pump to get water from the main canal in the field canals supplying the plot. Whereas high water levels at the beginning of Toman canal ensure excellent access to water by gravity irrigation, low water levels in the middle reach of the canal make it impossible to irrigate by gravity (Figure 4.12).

When some of the *nimra*s in the first reach complete irrigation a few days later and the first intakes are closed, more water becomes available for the tail end of the canal.[80] When more field canals close their intake, the water level rises so high that it starts to seep through and overtop the embankments at the tail end of the canal.[81] Yet while the water level is now high throughout the minor canal Talia still can not irrigate: her plot is the 21st out of the 30 plots along the field canal, and the seventh and the 8th plot are now irrigating for the 6th day in a row. Both the fact that field canal 15 has not been cleaned before cultivation[82] and the fact that the land around the middle of nimra 15 is high, limit

---

[80] This happened earlier than I expected from the limited gate opening: later I found that some of the cultivators at the tail end made holes in the last pipe regulator to make the water flow quicker to the tail end.

[81] This is not merely a matter of a phone call to the gate operator to close the intake of the minor: at the major canal level the problem replicates, or rather, is inversed. The engineer can not allow the sudden closure of minor canals as extremely high water levels are operated, and the sudden closure of one of the minor canals may jeopardize the banks of the major canals.

[82] The people did not pay for the machine which prepared the field canals which is owned by Elfatih.

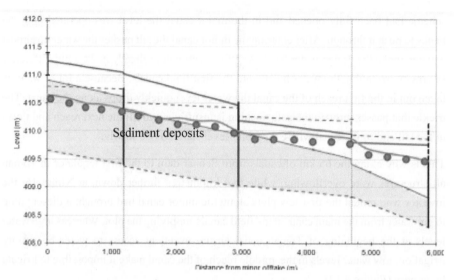

*Figure 4.12: Water level (blue) above the land level (green) along Toman minor canal on 24 September 2011. The deposited sediment is shaded in the figure. In the third reach of the canal (3000 - 4800 m from the offtake- nimras 12-15 are here) the water level in the canal is far below the Full Supply Level (dashed) and only just above the land to be irrigated (green). At the end of reach 3 the difference between the water level (Blue line) and the land (green line) is less than the 10 cm required for the water to flow by gravity. The only way to irrigate here is by using pumps. The blue circles indicate the positions of the 21 field outlet pipes through which water users take water from the canal.*

the rate of the flow (Figure 4.13). As it has been more than a month after the plots received water, fierce struggles erupt. As the informal rule is that the groundnut crop gets priority as it is more drought sensitive than sorghum, there was no way for Talia, who has no backing from Mahmoud in the field, to claim her turn. People accuse each other of taking too much water and try to jump the queue through closing the intake of irrigating plots at night and opening their own intakes. The land users at the tail end of the *nimra* face this problem every year and have started taking water through a *nacoosi*[83] pipe which taps into the neighboring minor canal which flows along the tail ends of the Toman field canals.

---

[83] A *nacoosi* is the diversion of water along a different route than was originally designed. Over the first decades of the scheme *nacoosi* field out pipes were added to the minor canals to be able to irrigate tail ends of the *nimra*s that were difficult to command because their land is higher. Since the 1980s 'illegal' *nacoosis* have been added. Three nacoosis from the adjacent neigbouring canal provided water to the tail ends of 12 out of 21 nimras in Toman (see e.g. the irrigation of *nimra* 12 and 14 through *nacoosis* in Figure 4.14).

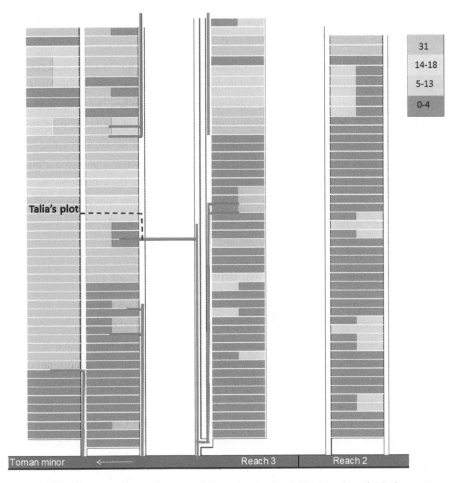

*Figure 4.13: Watering status of nimras 10-15 on 4 October 2011 based on field observations. Lines towards plots indicate active irrigation routes. The dotted line is the route Talia is digging. The numbers in the legend indicate the days the plots have been without water.*

This left Talia and the women cultivating surrounding sorghum plots in the middle part of the *nimra* in trouble. As it took too long to wait for the water to arrive through the field canal of their *nimra*, she started digging ditches to connect their field canals with the field canal of *nimra* 12 which was now overflowing. (Figure 4.1). Just before she finishes, the embankment of the minor canal breaks at the tail end and the level in the canal quickly drops. Many irrigators had closed their intakes and because water continued to flow into the minor canal, the water level at the tail end had again overtopped the embankment. Only after the breach is repaired can she access the water. On 11 October 2011, 38 days after the plot last received water from rain, it is finally Talia's turn. She irrigates both the groundnut and the sorghum plots in two days.

The performance of 'participatory water distribution', then, is not an ahistorical function of the ability of equally positioned community members to take control over their own water. The organisation of accessing water whenever it is available takes form through the regime of silt deposition and excavation, manipulations of weirs, mobilization of pumps and digging of new irrigation routes through which absentee tenants, tenant investors, migrant women, and contractors renegotiate their relative positions along the canal. The limits to participatory operation thus co-evolve with the canal form. As water flows through the canal abundantly but very irregularly, many along the canal grow drought tolerant crops which they supply with plenty of water whenever it is available.

## 4.7.    Dividing labour and sorghum

Talia does not feature in the plans of liberalization or IMT and we found her to be one of the last to get water. Yet this does not make her a hapless marginal producer. Strategically investing her labour in growing sorghum and groundnuts for own consumption or nearby markets, she and other women of Shazera manage to gain control over almost half of the cultivated area, most of it along the middle of the canal. The irregular supply of water makes them carefully spread their crops and labour over the land. From their particular position sorghum is the logical crop as it can be grown without fertilizers and pesticides. If the summer is not too dry, only one or two irrigations are sufficient for a good yield. As Talia only has to provide labour, she can tailor the intensity of the investment to the expected harvest. Because the final irrigation of her plot in *nimra* 15 was too late to get a decent harvest, Talia skipped the most labour intensive activity – weeding – and moved to plant a plot with *ladyfingers* in the neighbouring minor canal area. While this further reduced the yield of the plot in *nimra* 15, it optimized Talia's returns to labour.

Getting a grip on the wider pattern of yields along the canal is not straightforward. A large range in the planting dates and crop varieties leads to a variety in plant sizes and harvesting times. To get some sense of how location matters to yield along Toman canal, I mapped the sorghum harvest. Not only is sorghum the most prevalent crop; it is also harvested in a relatively short period around the first two weeks of November. Our first attempt to complement observations with interview data from tenants and sharecroppers was unsuccessful. Many of the tenants denied cultivation of poor yielding plots altogether, and this turned out to be about more than tenant pride or shame. The flat taxes for land and water that replaced the crop sharing agreement prove to play a crucial role here. The threshing machine - owned by Elfatih – only threshes crops upon handing over a certificate which proves that newly introduced land and water taxes have been paid. To avoid the costs of the threshing machine (10 SDG per bag) and the land and water tax (70 SDG per acre for sorghum), Talia and Mahmoud agreed that she would thresh the sorghum in Nimra 15 herself by hand. From the two acre plot the yield was just over five bags (0.5 ton/ha). Mahmoud took three bags because he provided the jute bags and will pay for land preparation and Talia took the remainder. From the 4 feddan plot *nimra* 12 she harvested almost 14 bags (0.7 ton/ha)[84]. The yields were less than halve of those of the plots owned and rented by Elfatihs in *nimra* 7, which varied between 1.8 and 2.1 tonnes per hectare.

As the information I collected from interviews and field observations was contradictory, it was impossible to develop a comprehensive overview of the 2011 sorghum harvest. I therefore decided to come back the next year to more intensively observe the harvest and threshing. While 2012 was not exceptional in terms of sorghum harvest, I found that the average yield was poor at around 0.8 ton per hectare. As wI expected from the differences I found in 2011 between the harvest of the plots of Mahmoud and Elfatih, the variation over the Toman command area was large. The harvest map (Figure 4.14) brings into view a distinct U-shaped pattern: yields at the head and tail were much higher than in the middle and yields near the canal were higher than those further away from it.

The map helps to understand how the implementation of irrigation management transfer does not lead to 'natural' selection of most entrepreneurial tenants by increased self-responsibility or a recovery of O&M costs by land- and water taxes. Instead, the differential positions for accessing land and water that are shaped with the canal grid has given rise to a diversity in forms of irrigation, cultivation and taxation. Whereas along the head and tail ends of the canal, which are receiving water first in periods of water shortage, many of those who came to be identified as 'Sudanese Arab tenants' rented land out to

---

[84] Despite his complaints of water shortage he pays 280 SDG tax. Right after the harvest he sold two sacks of sorghum to repay the short loan he took with a merchant to pay this tax. Further, Mahmoud paid one and a half sack of sorghum to the operator of the harvesting machine. The remaining 10 sacks were split between Mahmoud and Talia. Out of his five sacks Mahmoud sold two sacks to pay Elfatih for land preparation.

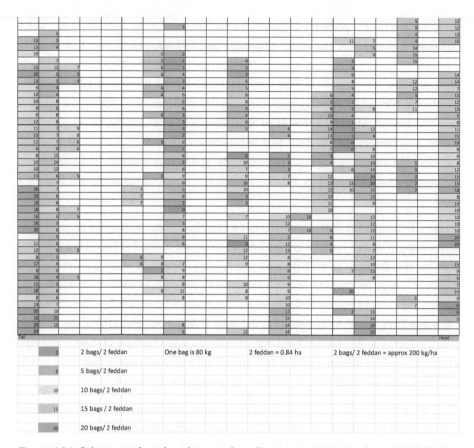

Figure 4.14: Schematic of sorghum harvest along Toman canal in November 2012 (left side is tail end. right side is head end!). Own observations of threshing November 2012. Note the U shape in yields along the canal.

their peers, who lent them money and machines and who hired male Arab *wakeel*s to supervise high input sharecropping by small groups of male 'migrants', on plots along the middle of the canal for which access to water was becoming increasingly uncertain, migrant women and children are cultivating larger plots of land with sorghum, and spread their water and labour wherever they expect the highest returns. In this way the recruitment of cheap labour, the expansion of cultivation of water, and – against the odds of many – the consumption of water in the Gezira and the production of food have been pushed up higher than ever before (Figure 4.15). The 'reformed' Gezira grid, then, provides the shifting frame through which people like Mahmoud, Elfatih and Talia continuously negotiate changing positions in the cultivation of irrigated plots in the Gezira.

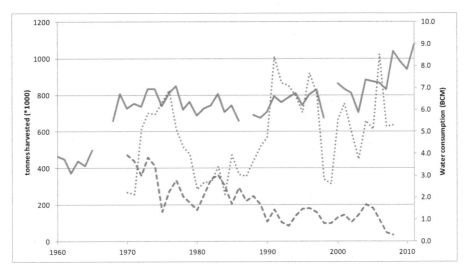

*Figure 4.15: Total water consumption and sorghum and cotton harvested in the Gezira scheme, 1960-2010. Water consumption in BCM per year (continuous blue line), Sorghum (dotted) and Cotton (dashed) in 1,000 tonnes per year. Sources: Tonnes harvested: Gezira scheme 2011, Water consumption: MoIWR (2012)*

## 4.8.    Conclusion

This chapter shows how the politics of irrigation reform in the Gezira materialize by analysing the changing relations between the physical shape of an irrigation canal and the class, race and gender positions of the people alongside it. Irrigation Management Transfer programmes across the Global South promote the transfer of operation management to water users as they are argued to be best positioned to balance production and risks. Yet most policy documents for irrigation reform have little to say about who those users are, or how they are linked by the canal. On the other hand, a rising number of scholars who point out how technological networks act as facilitators of coercion and commodification, often have equally little to say about the role of infrastructure and people in this process.

The analysis of practices of canal operation, maintenance, irrigation and harvest around Toman canal brings into view some of the limitations of seeing the scheme as an 'inefficient' system or merely a vehicle of capitalist exploitation. Focusing on the distribution of soil, water and labour, the chapter shows how repositioning of water users along the old colonial canal infrastructure reshapes both the forms and intensities of cultivation in the Gezira.

Despite the fact that more and more water is pushed into the Gezira to reduce shortages wherever and whenever possible, both excessive water supply and water shortages are on the rise in the Gezira. The rearrangement of canal operation and maintenance, and workers are crucial to understand this transformation. Recent reforms in the scheme operation and canal maintenance have made sediment removal a lucrative business. As the 44 private companies that are involved in desilting of canals today are paid by kilometres cleaned rather than by the quality of the excavation work, they leave the canal system covered with a thick layer of sediment. The reduced canal cross sections increase the differences in access to water along the canal. The increasing friction of the canal bed makes that water levels, and thus access to gravity irrigation water, have come to vary widely along the canal.

This also helps to explain why the most prevalent crop in the Gezira has changed from cotton to sorghum. Increasing fluctuation and uncertainty of the water level have played an important role in shifting to a crop which is less prone to drought and requires fewer inputs like fertilizers. The relative positions of the water users along the Gezira and subsequent terms and intensities of their engagement in cultivation, shifted with the transforming canal form and functions, and further strengthened class, race and gender differentiations that were shaped with the canal grid. With water levels operated as high as possible to supply ever higher cropping intensities, Toman canal was transformed from a wide and deep canal for storage of water over night to a narrow conveyance canal from which water is taken whenever it is needed and available. While at the head and tail ends of the canal migrant sharecroppers work for rich tenants who rent in land to invest in high yielding sorghum and vegetables, along the middle of the canal, where access of water is increasingly low and unpredictable, the hard work of cultivation has shifted increasingly from Arab male tenants and migrant male wage labourers to 'migrant' women and children sharecroppers with limited rights and poor access to drinking water, housing and education. They use hoes and gunny bags to divert the flow to fields on which they grow staple crops which are less water sensitive. While the scheme is thus devalued in the narrow terms of the formal economy recorded in books of government, hundreds of thousands of women and children use the increasingly unreliable water supply to expand the production of a staple crop for home consumption and for sale in markets where food prices are sharply rising.

The fact that the outcome of irrigation reforms in the Gezira are open to multiple interpretations – a state in decline, a vehicle for marginalization of labour, a vast territory up for cheap rents to grow high value vegetables for the Khartoum market, an opportunity for millions of migrants to earn from sorghum they sell - sheds a light both on its political nature and its different possible futures. Whether the emerging models of wheat, vegetable and sorghum cultivation in the Gezira are sustainable in the long run remains to be seen. With riots over the price of bread made of imported wheat intensifying (Sudan Tribune 2013), new forms of cooperation between tenants and migrants provided them with a possibility, through which they could retain a significant degree of control over land and water. The recent reengagement of the old irrigation bureaucracy under the ministry of water in the operation of Gezira canal provides another opportunity with prospects for a consolidation and improvement of food production in the Gezira. At the same time, however, increasing tensions over Nile waters are used by the government (GoS 2008) and the IMF (2014) to justify new investments in an 'old' model of export agriculture for economic growth, which might – as it will compete for funds and water with the Gezira - jeopardize the emerging model of domestic food production in colonial irrigation schemes. In the next chapter I look across the river, where a large part of a similar rectangular canal grid that was cultivated by smallholders has recently been replaced by large-scale centre-pivot irrigation, to see how the old hydraulic mission model takes new forms in the name of water productivity. I will focus particularly on the role of science and engineers in extending the lifetime of the hydraulic mission long after closure of the river basin.

# The (re)making of a water accounting culture: entanglements of water science and development in the Waha irrigation scheme of Sudan

## 5.1 Introduction

Soon after the construction of the Roseires Dam (1966) and the Aswan High dam (1970), the Nile River basin was 'closed' in the sense that almost all its waters were consumed before the river reached the Mediterranean Sea. A century of colonial and postcolonial irrigation development had reduced the Nile outflow into the Sea by some 85%. The remaining 15% was saline and therefore unsuitable for human, animal or plant consumption.[85] It thus seemed as if the hydraulic mission of not letting a single drop of fresh water of the Nile flow into the sea without being used beneficially by humankind had effectively been achieved. Yet, the Egyptian government decided that the mission could be extended even further: in the 1980s, under the heading of 'demand management', it engaged in a large-scale programme to optimize what it considered beneficial use: the construction of more efficient irrigation projects and plans to reduce the waste of water in existing schemes (MoI Egypt et al. 1980, Mohamed 2001).

Egypt's water demand management strategy fitted a global reframing from water as a resource freely available for human development to water as a scarce economic good. From the 1980s onwards, USAID (MoI 1982), the World Bank (1993), FAO (1993) and the ADB (2003), worked with governments around the world to turn water into a resource of which the supply and demand were to be regulated by the market, as this was believed to produce highest efficiencies. Doing this required new tools to not just trace the distribution of water, but to also attribute it with a market value. At the same time the calls by environmentalists and NGOs to account for the impacts of irrigation projects and dams on humans and the environment – impacts that had long been considered as external to hydraulic construction and development – became recognised (e.g. Rogers et al 1997). In the 1990s the UN took a leading role in drawing together these two emerging agendas: that of water as an economic good with that of water as a socio-ecological concern. At the UN conferences on Water and the Environment (Dublin 1992) and Environment and Development (Rio de Janeiro 1992), principles were agreed upon that firmly established water as a resource for global sustainable development. Instead of the nation state as the most important actor in developing and distributing water in the national interest, the new policy consensus was that the market is the most efficient mechanism for water allocation. By reducing water to terms of stocks and flow resources, governments and international development organisations enabled its subsumption in ecosystem services (Norgaard 2010). Furthermore, accountability for the organisation and allocation of water

---

[85] This does not mean it is it can be qualified as useless: it has a crucial function in halting further salt water intrusion.

increasingly became a matter of standardisation and transparency: much effort went into the production of indicators to measure the achievement of economic value produced through water development. These developments summarize in very broad strokes how a 'water accounting culture'[86] is taking hold of development organisations, academia, and agro-industry. Indeed Water Accounting, i.e. the use of remote sensing for "independent estimates of water flows, fluxes, stocks, consumption, and services" (wateraccounting.org – main page) is now widely perceived amongst international water development organisation as key to helping make the wisest water decisions, those that guarantee sustainable development of ever more uncertain flows of water resources (Molden 1997, FAO and World Water Council 2018, IWMI 2014).

The emergence of a distinct valuation and accounting culture in water has not gone uncriticised. Over the past three decades, criticism of the reification of water as a 'resource' or 'service' for environmental and sustainable development has increased. Scholars from a wide variety of disciplines have shown the limitations of narrowly depicting water as a resource and have questioned the possibility and desirability of the 'pricing' of ecosystem services in a market. An important thread in the different critiques is that expressing and deciding about water in market terms effectively obscures the hydrological, agricultural, and labour relations that make water what it is and that make its allocation possible (Savenije and Van der Zaag 2002, Domínguez Guzmán et al. 2016, Kaika 2005). The crisis rhetoric of absolute limits to water availabilities mobilized in calls for innovation and green growth has been answered by a growing literature which expresses deep concern about how such declarations of crisis legitimize undemocratic decisions that lend support to a 'global water grab' (Franco et al. 2013, Metha et al. 2012) of powerful entrepreneurs.

Both grand theories of 'green growth through frameworks of ecosystem services' and the 'global water grab' are targeted to policy makers with the aim to change and intervene. Yet they are not very useful for understanding the making of irregular patterns of water development along the Nile such as those in the previous chapters. In the Choke Mountains (chapter 3) and the Gezira Scheme (chapter 4) green growth theories do not help explain how large groups of people, who see increasingly more land and water diverted away from them, do not necessarily benefit from the realisation of higher water

---

[86] Naming the highly standardized set of procedures to account for the organisation of water a "water accounting culture" is inspired by the book "Audit Cultures", which provides several accounts about how accountability is increasingly equated with auditing in academia (Strathern 2000).

productivities. The suggestion of the water grab narrative that new forms of primitive accumulation are happening that lead to a seamless monoculture that feeds an undifferentiated proletariat is also not very helpful in describing the diverse experiences of people living with and alongside Nile waters. In chapter 2 I showed how the historical superimposition of dams on the river bed created highly uneven patterns of water distribution and use along the Nile. In chapter 3 and 4 I took a closer look at the making of state institutions and hydraulic infrastructure to argue that their power and workings should not be taken for granted but be understood as embodiments of continuous struggles over the organisation of Nile waters. This chapter discusses the question whether this does not equally apply to the changing categories of science and engineering used and mobilized to make the Nile visible and render it governable. Are the categories and conceptual languages of science and engineering not themselves part of the struggles over distributing Nile water? And if this is so, how am I, as a scholar of Nile waters, implicated in the making of the Nile water patterns that I study? How to position myself in an academy in which an audit and accounting culture is steadfastly taking hold?

This chapter addresses these questions by analysing how knowledges of water accounting now thriving in my own academic department relate to the highly particular patterns of water distribution they visualise. The analysis focuses on the politics, achievements and limits of what Donna Haraway calls the 'god trick': the move by which modern scientists place themselves and their instruments outside the making of the objects they study (Haraway 1991). Scientists and policy makers involved in 'water accounting' tend to insist that their procedures (e.g. scripts to transform satellite imagery into 'water accounting' maps) are neutral and their categories (e.g. coefficients of water consumption and biomass production) natural. Yet, it is not very difficult to see how their water accounting procedures produce a very specific type of water - water as a resource for the provision of goods and services – and in doing so render these legible for calculation and appropriation by policy makers, development banks and investors in ways that align to specific political and societal projects (Linton 2010). Indeed, water accounting maps represent (and produce) the world as consisting of a global population differentiated by degrees or levels of water productivity. Such maps feed into and build on a notion of development as consisting of increasing crops per drop. Yet land- or waterscapes always exceed what can be (re-)presented in water accounting maps, which means both that other representations are possible and that no single representation will ever fully contain real-world complexities.

Inspired by Tsing's analysis of what happened when different commitments to 'nature conservation' met and mixed in the Meratus Mountains of Indonesia (2005 - Chapter 6), this chapter analyses how water scientists and water users engage and relate in the making of 'water development' around the Waha centre-pivot irrigation farm along the Blue Nile in Sudan. Following Tsing's example, I re-trace multiple 'scientific' and 'not so scientific' accounts of hydraulic development and their relations. In this way, this chapter unravels how scientists and water users take part in and are jointly responsible for the production of highly particular knowledges and patterns of irrigation at Waha. By appreciating water accounting as a historical and material process, the chapter aims to shed a different light on its potential and limitations for rearranging the organisation of Nile waters.

To ground the analysis of the politics of water accounting in this chapter, it starts in two places that were chosen because of how they present incidences of overlapping multiple modernities of water science and development (section 5.2). In the area that came to be known as Waha along the Blue Nile in Sudan, recession irrigation – i.e. cultivation using the water left on/in the banks of the Nile by the receding floods – was complemented by a cooperative smallholder irrigation scheme, of which most was recently transformed to a private farm using centre-pivot irrigation (Figure 5.1). This change somehow parallels a gradual change in my own organisation IHE Delft Institute for Water Education (initiated in 1957 with funds of the World Health Organisation and the Organisation for Economic Cooperation and Development) where a culture of direct technical support to postcolonial governments to strengthen their capacity in the fields of hydraulics and environmental engineering is steadfastly being replaced with a culture of producing ever more accurate systems to measure and account for sustainable water development[87].

Section 5.3 analyses how earlier projects of modern hydraulic development in Sudan conditioned the positions of scientists, productive and unproductive users in the making of what I call the water accounting culture[88].

Sections 5.4 to 5.6 map the shaping of both the forms and meanings of water productivity at Waha today. This mapping brings to the fore why the modern ideal of water security/productivity, today formulated as 'using water to feed the community of Sudanese people and the world', remains once again unfulfilled as a new high tech

---

[87] For a prominent example of indicators addressed in this culture see UN Water (2016)

[88] Naming the highly standardized set of procedures to account for the organisation of water a "water accounting culture" is inspired by the book "Audit Cultures", which provides several accounts about how accountability is increasingly equated with auditing in academia (Strathern 2000).

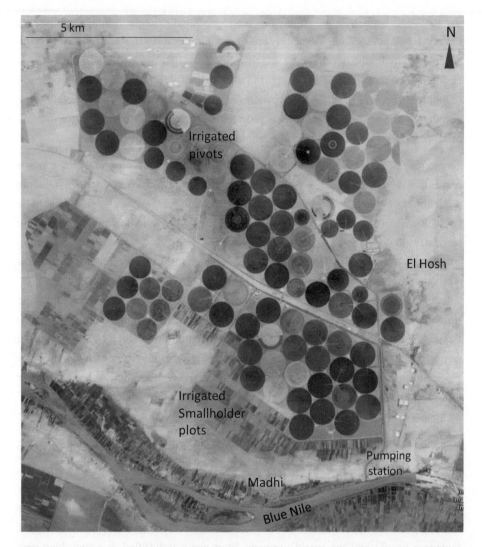

*Figure 5.1: Waha farm and surroundings. Some of the plots near the river are irrigated by recession irrigation after the receding Nile flood. The pumping station provides water to both the remaining smallholder plots and the private centre-pivot irrigation farm that was largely built on top of it. Source Google Earth 2016*

irrigation scheme replaces part of the old smallholder scheme. The sections also show how these emerging engagements reshape differences in 'communities' of scientists, productive and unproductive farmers and their possibilities to rearrange the distribution of water at Waha.

Bringing 'water accounting' maps produced in the ministry and my own academic department in conversation with accounts of the water users at Waha –those who are rendered as more or less productive by the maps – shows how water accounting maps and irrigation schemes take form together in the water accounting culture. Studying how the particular geometries of scientific knowledge and infrastructure transform through multiple contested and overlapping projects of knowing and engineering, opens possibilities to change the water accounting culture in ways that do not concentrate the benefits of using Nile waters only for those who the water accounting maps recognizes as productive.

## 5.2 Overlapping geometries of water control – engaging with Waha and Integrated Water Systems and Governance (IWSG)

AGA's 7,000 ha irrigated Waha farm is located right across the Blue Nile from the smallholder Gezira scheme discussed in the previous chapter (Figure 5.1). When driving on the new highway from Khartoum to Wad Medani on the east bank of the Nile, sparse vegetation suddenly makes place for densely cultivated circles of alfalfa supplied by water 24/7. Closer towards the river, the circular grid abruptly changes again to the rectangular plots similar to the ones in the Gezira (Chapter 4). What makes these distinct geometric patterns?

To answer this question, many at Waha refer to a series of events that took place between 2005 and 2008. In 2005, government forces were sent to Waha to put a fence around the irrigation scheme in which agricultural production had declined and for which the government no longer paid the maintenance. The land was leased to AGA Company. Soon after, the company's bulldozers demolished some 80% of the existing rectangular irrigation grid to construct new minor canals and equipment for centre-pivot irrigation. The ends of the new minor canals were reconnected to the tail ends of the old irrigation scheme. Average annual biomass growth in the new scheme is now some 15 times higher than production on the remaining irrigated smallholder plots (that are now supplied by AGA Company water) (HRC 2016). Unlike the earlier scheme which was designed for supplementary irrigation, the new scheme does not rely on rainfall and produces alfalfa for export to Saudi Arabia. This fodder crop is on the field around the year. By 2015 the AGA farm had become the new model for agriculture promoted not only by the company itself but also by the Sudanese government, the IMF, Dutch, British and US embassies and a range of other government officials that visited the farm (BBC 2013, FT 2014).

It was this aura of success which drew me to Waha. The project was pointed out by my Sudanese colleague Professor Osman when brainstorming about a grant proposal for a research project on the redistribution of water through large-scale irrigation development along the Nile. He described how over the past decade AGA had managed to set up a brand new irrigated farm while other modern irrigation projects were faltering or not forthcoming. Apparently, AGA was very successful in establishing a new model for irrigated agriculture. This was all the more remarkable because the project was designed by some of the same Sudanese engineers who had left Sudan in the 1980/1990s, after the Arab financed projects in Sudan they worked on had largely failed to materialize (Woertz 2013). In Saudi Arabia they had worked on implementing the Gulf country's new strategy for high tech production of irrigated crops within its own borders. Many of these engineers were now returning.

There was also a more intimate connection which drew us to Waha. The distinct geometries of cultivation at AGA had evolved in tandem with the type of project work that the water institutes that employed Osman and myself were asked to do by their funders. Confronted with foreign debt, the outbreak of famine and the collapse of the Numeri regime in 1985, the Bretton Woods institutions had halted project loans for support to large irrigation schemes. Instead, they offered policy support in exchange for structural adjustment: in exchange for downsizing state expenditures, currency devaluation and a reorientation of state support from publicly managed cotton schemes to the private sector, the UN, NGOs and water institutes became increasingly involved and indeed gained control over tasks in agricultural research, food distribution and social welfare that were formerly carried out by the state. This was the context in which the Ministry of Water Resources and Irrigation, still powerful in the 1980s, became the subject of three decades of austerity measures. And this was the setting in which Osman received a scholarship to do his PhD studies at IHE Delft in the Netherlands. When he completed his PhD in 2005, IHE had turned into the UNESCO institute for Water Education. Meanwhile, the staff and influence of the irrigation ministry of Sudan had steadily diminished (GoS 2005). Old state irrigation schemes were transferred to Water User organisations (Chapter 4) and new private irrigation schemes like AGA's farm at Waha operated almost entirely outside of ministerial control and support. Frustrated about the neglect of irrigation agriculture but refusing to leave Sudan, Osman returned to his home country to contribute to the resuscitation of the fame of the Hydraulic Research Institute in the irrigation ministry. Facing the reality of working in a ministry with declining power and money, he successfully applied for a part time position as an associate professor at UNESCO-IHE to forge connections and funding beyond Sudan's borders.

In November 2014 Osman and I sat down over lunch at IHE to discuss our application for a research grant in response to a call of the Consultative Group of International

Research (CGIAR). CGIAR is a "global partnership that unites organizations engaged in research for a food secure future" (CGIAR 2011). The partnership had evolved in the 1970s from the rice, wheat and maize improvement institutes that were instrumental in globalizing the green revolution. When the limits to the green and irrigation revolution came in sight (in the 2000s), the CGIAR reinvented itself by pro-actively inviting civil society and the private sector to join in the global search for green growth and inclusive sustainable development, thereby broadening its portfolio to bio-engineering and the sustainable management of natural resources (CGIAR 2011, p5). This was the context in which CGIAR's programme on Water Land and Ecosystems issued a call for expressions of interest "seeking to develop a portfolio of innovative complementary projects that will promote agricultural and related investments, in a framework of sustainable ecosystem services that equitably meet the needs of women and men (food security, nutrition, higher incomes), while also promoting economic growth" (CGIAR 2014, p1). The CGIAR was authoritative and confident on how to best do this. It called for an ecosystem-based approach which it defines as: "Harnessing ecosystem services for production goals (e.g. increased yields, nutrition, and goods) or in ways that support these goals (e.g. controlling pests, diseases and weeds, regulation of hydrological flows), while reducing negative impacts on the natural resource base providing these ecosystem services." (Ibid, p2).

This prescription of a framework for the valuation of environmental services in support of combined environmental and livelihood improvements was simultaneously uncomfortable and familiar to both Osman and myself. We were worried about the apparent widening of the gap between the rhetoric of development by increasing stakeholder cooperation and building climate resilience (NBI 2014, UNEP et al. 2014) on the one hand, and the harsh realities of food riots, state neglect of irrigation systems and unilateral decisions to construct new dams on the other (Sudan Tribune 2015, Zenawi 2011). Our discomfort also stemmed from a realization that calls for proposals like the one by the CGIAR provide our bread and butter. Indeed, our academic department had slowly transformed with the changing funding opportunities for water research and education. Since 2012, twelve new staff members with expertise in 'global water accounting', 'global water-energy-food assessment' and 'water diplomacy' were hired in our Integrated Water Systems and Governance (IWSG) department.

These new positions fitted well with the framework proposed by CGIAR "to bring ecosystem services to the forefront of the global development agenda" by "measure[ing] performance" (CGIAR 2014, p4). Yet with their rendering of development in terms of "economic efficiency", and "good governance", the relations between researchers and water users were restructured in ways that posed key challenges to the engagement of researchers in local water development practices. First, as the expertise of our department was increasingly geared towards remote sensing, global modelling of water food flows and diplomacy, the literal distance between the concerns of these water scientists and those of the entrepreneurs involved in the construction of hydraulic infrastructure had widened considerably. Second, as the same accounting culture had rendered irrigation in Sudan an almost exclusively private sector matter, our scientific involvement in irrigation was only possible if we succeeded in somehow convincing a company we would work with that it could benefit from it.

This was the financial, institutional and political reality in which Osman and I strategized about pursuing our own political scientific objectives. For us, the new model of large scale private farming by AGA at Waha provided an interesting yet challenging entry point to engage with the ever more contested distribution of Nile waters and the lure of technological advancement. We assumed that AGA would be interested in remote sensing based imageries that showed how water, biomass and nitrogen were distributed over their farm, as they could perhaps use these to improve their farm operations. Enticing them with this prospect, we asked them the return favour of being allowed to do interviews in and around the farm to provide a historical analysis of changing relations of water, land and work in Waha. Osman and I shared an interest in analysing the ways in which the profoundly different accounts this would yield and how these could be related, contrasted and compared. We engaged with scientists from the University of Khartoum and Sudan's Hydraulic Research Center to formulate a project with the objective: "to explore how remote sensing tools and ethnographic methods could be used and combined to understand how large scale irrigation development at Waha re-configures relations of land, water, labor and food, and what this means for equity and sustainability" (UNESCO-IHE et al 2014, p3). For Osman, the project provided an opportunity for the Ministry to re-establish its influence and reputation in Sudanese irrigation development and contribute towards reviving the productivity of Sudanese irrigated agriculture. My own interest was to explore relations between the remotely sensed 'water accounts' that were becoming increasingly popular in our academic department and the actual practices of water diversion from which they seemed ever further removed.

The following sections provide my interpretation of the relations between the different accounts of water produced by the project. To do so the next section further explores how the rise of the audit culture in global water management has been both a product of, and instrumental to, a re-appropriation of the lands and waters that used to be committed to colonial and postcolonial projects and bureaucracies in Sudan. The sections thereafter bring into view how different accounts of water development at AGA's farm in Waha and its surroundings are both forming and structured by this water accounting culture.

## 5.3 Reinventing the bread basket in Sudan - Rise of audit culture in Nile water management and its effects

The notions of efficiency and productivity that are central to contemporary accounts of modern hydraulic development have a lot in common with the scientific categories of mathematics, hydraulics and hydrology that were mobilized by British engineers to help justify colonial control over river basin in the 19th and early 20th centuries. This section shows that in spite of historical similarities and resonances, today's categories of modern hydraulic development are not the same as those of colonial times. Rather, 'science for development' has transformed with the changing relations of land, water and labor it seeks to legitimize. In the process, the respective roles of water, its users, scientists and policy makers also have changed.

In the late 19th century, statistical knowledge of flood occurrence, experimental farms and 'natural' boundaries of river basin were considered the universal scientific basis for establishing the potential for, and implementing projects towards, the 'betterment of native populations' (Churchill 1899, Willcocks 1904). This knowledge was produced and used by ministries of water and irrigation that were being established around the world. Along the Nile, British rulers and investors envisioned 'development' as happening through modern cotton cultivation. Securing a reliable supply of water was central to this project, which required the control over and storage of water in dams in the upper basin. Scientific maps of the river basin, hydrographs and advances in storage dam construction thus became key objects in the legitimization of the British conquest of Sudan in 1898 (Tvedt 2004).

Like elsewhere in the British Empire the violent appropriation of land and the forced production of cotton and sugar for the 'global market' produced its own reversals. Massive landlessness that accompanied the conversion of agricultural lands to sugar and cotton estates fuelled nationalist movements around what came to be called the Third World. Increasing European and American challenges to British hegemony in world trade prefigured the 1929 financial crisis and the downfall of British world hegemony. To protect landed interests in modern development, the administration of trade and development was rearranged along the lines of national economies, national currencies and national populations. This was the context in which the US and European states set up the Bretton Woods Institutions (1944) which would, after the second World War, be mobilized to finance megaprojects for continued production of tropical commodities like cotton and sugar, while also leveraging Western (and Eastern) influence in the cold war era. Nation states divided rivers into national shares of water. While the subsequent construction of gigantic hydraulic projects was presented by governments as a process of national development, this process profoundly rearranged ecological relations and created new international dependencies (Sneddon and Fox 2011, Ahlers et al. 2014, Sneddon 2015). State led mega irrigation projects that were constructed in the South in the post WWII era relied heavily on synthetic fertilizers, pesticides and machinery produced in the North, and large amounts of water. As rivers were depleted for the production of cotton and sugar, the increasing substitution of these products by synthetic fibres and sweeteners undermined their markets. While the involvement of African farmers in the production of global tropical commodities created a growing market for subsidized wheat producers of the North, the promised trickling down of benefits of high value export production to the same African farmers never materialized. From the late 1970s, it became increasingly clear that mega-dams of the post-World War 2 era had not only boosted the production of global commodities and widened global inequalities, but also sparked environmental and economic crises around the Global South (Chapter 2, WCD 2000).

With land, water and labor increasingly committed to colonial and postcolonial projects, the IMF, World Bank and the Arab Fund discontinued the provision of loans for the construction of public irrigation projects as their main strategy for Sudanese development from the 1970s (Ali 1989, World Bank 1982). The World Bank gave the high public costs of growing irrigation bureaucracies and the approaching limits to the full utilization of the Nile as agreed by Egypt and Sudan as key reasons (IBRD 1982). Instead, it worked with the government and the Arab Fund to identify the facilitation of foreign private investment in smaller but capital intensive irrigation schemes and large scale mechanisation of rain fed land as their priorities in Sudan.

Keen to reinvest oil money at a time of global crisis in which opportunities were limited, investors from the gulf dubbed Sudan as the 'bread basket' that would provide food-security for the Middle East in the aftermath of the Arab-Israeli wars. Until the 1980s, almost 4 million hectares of rain fed land would be turned into privately mechanised farmland using Arab capital and western technology (Duffield 1990). Yet the productivity of most of these farms dropped dramatically within a few years after the land had been cleared (World Bank 1979). With soils mined and substantial food exports never materializing, only those involved in financing and facilitating the bread basket investment benefited from it. The influence of Arab investments in Sudanese agriculture persists until today. When oil and food prices started to pick up in the new millennium, the bread basket was reinvented for irrigated agriculture. Since 2003, millions of hectares of land along the Nile were leased to companies from Qatar, Jordan, the UAE and Saudi Arabia (Woertz 2013, Sandström 2016). As highlighted by Woertz many of the leaseholders have not used these lands until now (Woertz 2013). They should, therefore, be interpreted as a hedge against further rises in the food markets, rather than as productive investments.

In the present conjuncture of new large investments in hydraulic development (2000 - onwards), the discourse of closing river basins has muted. It seems to have been replaced with a proclamation that it is imperative to produce as many high value crops as possible out of ever more uncertain resource flows. The focus on crops per drop has become so pervasive in policy, media, and academia that the fact that water is a resource for sustainable development is increasingly taken for granted. In the process, the roles assigned by government officials and international organisations, users, policy makers and scientists in water have been profoundly rearranged:

> **Water** is no longer regarded as free natural bounty to be harnessed for development. Neither is the state still imagined as the entity to define what (water) development should look like. Instead, in the face of depleting rivers and massive public debt around the world, a key objective of international and national investment and knowledge development efforts has become the resilient adaptation to increasingly uncertain global food and financial markets in order to achieve sustainable development goals. With water scripted as a scarce resource for global development, new opportunities for investment are opened up for global/export agriculture (GoS 2008, IMF 2014).

> **Water users** are increasingly rendered as entrepreneurs whose success depend on their ability to adapt to an ever more uncertain environment (GoS 2005). In the neo-liberal logic, people are no longer regarded by the irrigation bureaucracy as members of a relatively passive population subject to "observation, taxonomy and classification" for moral and national development (El Shakry 2007). Instead, water users are identified as best placed to distribute resources efficiently, and only those with the ingenuity to increase "nature based income and culturing resilience" (UNDP et al 2008 cited by Watts 2011, p lxxxi) are identified as fit for survival.

In line with the above development, **policy makers and state engineers** are no longer regarded as the managers of basins or planners of development for the Sudanese population. Instead, their tasks have been rescripted in two ways. First and foremost, they are expected to facilitate a transparent allocation of resources. For water, this takes the form of keeping account of its distribution and of the benefits it generates (GoS 2005). This role has considerable overlaps with the roles of **consultants and scientists**, which take over ever more work from the Sudanese state bureaucracy. Second, government officials work with development organisations and NGO's to provide so-called safety nets for people who fail to pass the modern test of resilience (World Bank 2018a, FAO 2016).

The water accounting culture thus created brings into being two separations – one between productive and unproductive water use(r)s, and one between the facts of water productivity and the experts who measure them. These separations are not 'just' a particular scientific way representing water, but are also actively institutionalising a particular way of doing water and governing (controlling) it. The first separation establishes accountability in highly particular terms of productivity. The second separation establishes water accounting facts as 'truths' which can be independently known (cf Latour 1993, Zwarteveen et al. 2018). The categories and roles assigned to nature (land and water), water users, policy makers, scientists and NGOs in the water accounting culture, have been used to legitimize redistributions of land and water used by colonial schemes that were no longer productive to the Sudanese regime or the Northern institutions that sponsored this research. This has helped the Sudanese regime to realize a new model of agriculture – one that relies less on the irrigation bureaucracy and tenant union that had been instrumental in overthrowing governments in 1970 and 1985 and one that, through an exemption of international boycott for money flows for investment in agricultural production, offered ways circulate part of its income from oil (Verhoeven 2015). For the World Bank, the IMF and the Arab Fund this rearrangement leverages their influence over a political regime that increasingly has come to be regarded as a security threat, while also offering an opportunity for expanding food sovereignty beyond its own territory (Dixon 2014, Verhoeven 2015). To entrepreneurs like AGA the accounting culture provides a way to access new tracts of land and water use concessions.

As the water and land around the Nile is ever more invested with overlapping commitments, infrastructures and institutions, the re-visioning of agriculture as consisting of the optimization of crops per drop does not proceed unimpeded. Although the making of 'objective water accounts' is actively promoted by IWMI, UN's FAO and my own institute IHE Delft, the 'win-win solutions' through increases in water efficiency that often accompany these accounts are almost always slow to materialize and generate their own new limitations. The next section turns to contrasting accounts of water distribution in Waha to make visible how the clear boundaries between AGA's privately irrigated farm and surrounding lands came about through protracted and on-going struggles over land, water, food and knowledge. Neither water accounting scientists, nor the users they identify as productive or unproductive, adhere to the roles (of making water flows transparent and producing for export markets respectively) foreseen for them in NBI, IMF, World Bank, FAO or government reports on the distribution of Nile waters. How they engage with the making of water productivity at Waha reveals the emerging limits to the water accounting culture as well as the new productive opportunities that surface with these limits for food producers and consumers along the Nile.

## 5.4. Water accounts as mappings of potential for development

It was our own research project which introduced the idea of water accounting to the AGA management. On 9 March 2015 our team received the technical manager and the farm manager of AGA's Waha farm at the Hydraulics Research Center of the Ministry of Irrigation and Water Resources. This was not just to 'tick the box' of involving 'the private sector'. As the ministry was no longer involved in irrigation at Waha, we needed the support from AGA to be allowed to study Sudan's new model of irrigated agriculture.

Osman's colleague Nour presented the part of the project which was most interesting to AGA's managers. In a presentation called 'The Possibilities and Impossibilities of Remote Sensing for Inclusive Water Accounting' he showed the impressive draft results of a quick processing of satellite images which displayed values of biomass growth, water evaporated and transpired, and water productivity (biomass/ET) for every 30 m x 30 m grid cell of the area in which the AGA farm is located. On the map, AGA's Waha farm lighted up as if it was an Oasis (*Al Waha – Arabic*) of water productivity (Figure 5.2). Nour explained that when calibrated with the water consumption per irrigated area, these images could be used to improve productivity by better adjusting water supplied to the plots to crop water requirements. In the context of a river increasingly committed and flows ever more uncertain, he went on, water accounting is becoming key to rational development of Sudan's water resources.

Water accounting allows for and helps producing a new form of governance of Sudan's water and its people. Accounting for the organisation of water based on remote sensing makes water and is users subject to what Strathern (after Douglas 1992) calls specific "rituals of verification from the outside" (Strathern 2000, p3). Water departments of universities and water ministries around the world have become institutions in which calculations of efficiency and principles of transparent governance are becoming

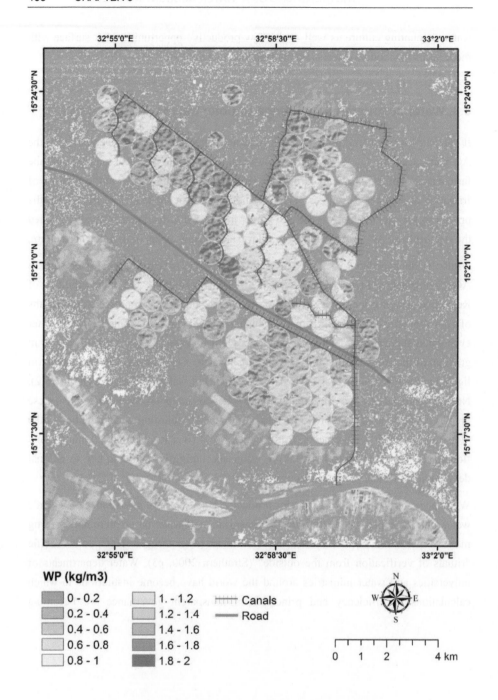

**WP (kg/m3)**

| | | |
|---|---|---|
| ◼ 0 - 0.2 | ◻ 1. - 1.2 | ┼┼┼┼ Canals |
| ◼ 0.2 - 0.4 | ◻ 1.2 - 1.4 | ▬▬ Road |
| ◻ 0.4 - 0.6 | ◻ 1.4 - 1.6 | |
| ◻ 0.6 - 0.8 | ◼ 1.6 - 1.8 | |
| ◻ 0.8 - 1 | ◼ 1.8 - 2 | |

0   1   2        4 km

*Figure 5.2: Water productivity for the Waha area, April 2014 – March 2015.*

pervasive indicators of their own good governance. As government officials like Osman were forced through a number of irrigation reforms to retreat from irrigation planning and management, water ministries are increasingly expected to measure the performance of irrigation schemes in terms of universal indicators (like crops per drop, or US$ per drop) that enable an assessment and comparison of individual farmers. The roles of government officials and scientists thus merge in a joint production (and performance) of transparency in the distribution of water ('data democracy'), a process in which they also re-position themselves as the stewards of water (Wateraccounting.org). The logic is that by making all use(r)s comparable vis à vis each other (in this case through the use of indicators of biomass production, water consumption and water productivity, i.e. kg biomass/m$^3$) more efficient ways to do water distribution can be imagined and implemented (water can be (re)distributed to those who are most productive/efficient). The logic thus re-makes water users into independent individuals whose productivity and performance is (and can be) constantly measured both for their own enhancement as well as for a more abstract collective good: "sustainable development". Judging from the hard facts of biomass production displayed on the map presented by Nour, there was only one farmer who passed the test of good water stewardship. Not by coincidence this was the farmer present at our workshop: the representative of Waha farm.

Osman and Nour are well aware of the peculiarities of the Water Accounting scripting of water productivity. The script makes it relatively easy to quickly produce draft maps of the distribution of evapotranspiration and biomass over large areas with the use of publicly available satellite data. Yet, especially the rainfed and smallholder irrigated crops outside the AGA's farm have difficulties to make it to such maps in an accurate fashion: these plots rely for at least part of their water supply on rainfall between June and September, which is also when the presence of clouds inhibits clear satellite pictures.

But even when the model is further refined and standardized to deal with clouds and smallholder plots, Osman is well aware that the map hides more about entrepreneurship than it reveals. As he is himself the owner of a smallholder farm in an irrigation scheme near Khartoum, he knows that increasing crop productivity is more than merely a matter of water, soil and human ingenuity. To set up a farm such as Waha in Sudan it is not enough to access land and water along the Nile. The high investment mode of farming also relies on foreign capital, technology and markets that are heavily constrained by the prohibitions on trade transactions with Sudan imposed by the United States since 1997 (US 1997, 2006, 2017). Osman is therefore quick to admit that the logics of remote sensing science or environmental accounting that underlie the Water Accounting framework are of limited value for understanding how AGA managed to become successful in accessing land, water, capital and the technologies that were key to its farming success in Waha.

Despite these reservations, Nour and Osman insist in subscribing to the neutrality of the water accounting maps they produced. Even though well aware that thus bringing into view water productivity helps foreground AGA's farm as a model for success, they do not buy into the idea that water accounting is a selective or biased exercise. Water accounting, they hope and contend, will in the end benefit both AGA and the smallholder farmers around it. Their argument is that better insight into the distribution of water and plant growth will make water savings possible, which in the end will produce more water to be shared among contending users.

On the one hand, the water accounting culture presents the mapped distribution of water productivity at Waha as a neutral account of water that will *facilitate* users to optimize their flows. On the other hand, by doing so, practices of Water Accounting of Waha order people in highly specific categories of 'productive' and 'unproductive'. The account of a yield gap thus produced – i.e. the gap between the actual yields and the potential yields under optimal practices of cultivation - resonates with familiar stories of empty lands to be brought into development (Ertsen 2006) and a potential Sudanese bread basket to feed the Arab world (Woertz 2013). This time the vocabulary is not of a civilizing mission, or national welfare, but of green global growth and resilience.

One could argue that thus investing hope and trust in Water Accounting implies turning a blind eye to the violence and politics that characterized earlier projects of water development. For Osman and Nour, on the contrary, it is a conscious way to re-engage with these politics, politics that increasingly frustrate and elude them. With a constituency in finance, oil and real estate (El-Battahani 1999, El-Battahani and Woodward 2013), the government that holds power in Sudan since 1989 hardly cared about or reinvested in irrigation. Soon after the once powerful irrigation department had been downsized, security officials took over the main building of the Ministry of Water Resources and Irrigation and hired German and Australian consultants to design hydropower dams for them (Chapter 2 this volume, Verhoeven 2015). Meanwhile almost 6 million Sudanese have become food insecure (WFP 2019). In this context, for Osman and Nour, water accounting provides instruments to refocus political attention on the production of food and the distribution of water. For them the high yields of AGA's Waha farm provide hope at a time in which Sudanese wheat imports have grown to over 2 Mton/year (FAO 2018).

Osman and Nour's strategy is not just wishful thinking. They know that despite all claims to the neutrality of maps, scientific projects like our own have not left irrigation at Waha unchanged. For them, mobilizing water accounting increases the convincing power they have within the ministry to justify new public reinvestments in irrigation. They know too that these investments will not change the landscape in precisely the ways imagined by the scientists and engineers that do the mapping. As the following account of AGA shows, our project contributed to tying Waha to a host of new networks and actors by which irrigation at Waha is rearranged. Nevertheless, seemingly 'marginal producers' are not just bystanders that are helped by or compensated for these new configurations. They take active roles in creating the water accounting culture and its limitations.

## 5.5 Water accounts as global sustainable development in the making

Two days after our opening workshop, the farm managers of AGA received us to proudly show their reality of Waha. Our story and the initial maps of AGA's high productivity farming at Waha had also captured the interest of the UN's Food and Agriculture Organisation and the Embassy of the Netherlands. They were keen to join our project trip. As if AGA did not want to waste time in putting their slogan of 'feeding the world' into practice, our six car convoy was welcomed at the farm with well-filled food boxes. The AGA farm managers made sure we would not visit the farm with empty stomachs.

It was this same promise of feeding the world that had drawn AGA to the Waha farm in 1990. At the time, its subsidiary SURAC (Sudan Tractors Company) was requested to level and plow the land for the Three Capitals Cooperative Society. Abdel Gaffer Ali, a Nubian businessman after which the AGA Company would be named, had laid the foundation for his business empire when he founded SUTRAC and became Sudan's agent for Caterpillar tractors in the 1966. Between that year and 1990, when the tractor company was awarded the contract to plow the land at Waha, AGA had grown into a major player in the importing of equipment and executing land preparation in the Gezira. Yet at that point in time nobody could have known that 25 years later AGA would itself lease the land and become its farmer.

In the years after 1990 under de leadership of Abdel Gaffer Ali's son Ahmed, AGA continued expanding its services. The company had started trading the wheat which it received as part of the payment for Gezira land preparations. Even without associating with inner government circles, AGA was successful in making itself indispensable to the government. By exploiting its ties to foreign export companies, AGA managed to become Sudan's largest trader of staple crops (FT 2014, Mann 2013). With the construction of Sudan's largest flour mill (1996) the company earned the country's major concession to import wheat. AGA's flour imports were exempted from the US prohibitions to engage in trade transactions with Sudan in 1997 (US 1997), allowing AGA to become Sudan's most wealthy and powerful private business enterprise. AGA tapped into a thriving 'market': The broken Bread Basket dream had not only left Sudan with a public debt of 45 Billion USD (IMF 2014) but also with millions of people displaced from their land and dependent on imported (first mostly US produced and later mostly Australian) wheat. By 2000, Sudan had become Africa's third largest grain importer (after Egypt and Nigeria). While the domestic production of wheat remained stable at 1 Mton/year since 1995, the urban population relying on subsidized bread grew steadily with imports rising from 0.5 Mton/year in the 1995 to 2.3 Mton/year twenty years later (FAO 2015). In 2014 AGA controlled more than 50% of Sudan's wheat imports (AGA Group cited by FT 2014). As the owner of the country's largest flour mill, AGA became a strategic ally not only of international companies, but also of the Sudanese government, which depended on AGA to get wheat imported into Sudan. The government subsidizes 'bread' by allowing AGA to buy wheat abroad at an exchange rate far below the official rate (Siddig and Grethe 2015) and demanding AGA to provide the flour to bakeries at fixed prices throughout Sudan.

Around the turn of the millennium, AGA thrived from increasing wheat imports. Seeking to further exploit its exclusive ties to both government and foreign companies, AGA decided to expand its enterprise to become Sudan's major agricultural producer. According to the president of the Three Capitals Society that was formally in charge of farming land at Waha in the 1990s, AGA struck a deal with the governor of Khartoum and the Kuwaiti owner of the pumping station to start farming for export in Waha in 2008. The Kuwaiti owner agreed that AGA would pay 1.5 million USD for the pump and main canals. The government - since it started exporting oil (1999) no longer economically dependent on agriculture - provided the company with a 25 year lease of the land and 'free water'. Encouraged by the government and the IMF seeking to diversify Sudanese exports - and careful not to compete with its own subsidized wheat import business - AGA started producing fodder for export to Saudi Arabia.

AGA's farm manager is conscious and frustrated about the seemingly awkward situation in which AGA sells fodder fed to air-conditioned cows abroad while millions in Sudan rely on food aid. He resolutely points at the government: "We can feed the world, but not ourselves [Sudan]... The government is neither providing farmers with expertise nor helping sustain a suitable business environment when it comes to agriculture." During our tour of the farm, the managers showed us how the Alfalfa grown at the farm is irrigated, harvested and piled up for export for consumption by cows in Saudi Arabia. On hot days, the pivots supplying water to the circles with Alfalfa are running 24/7 (Figure 5.3). In the office, one of AGA's managers explains to us how this is only the beginning. According to the farm manager, Sudan's land is still vastly underutilized because there is no electricity to pump the water. At Abu Hamad, further along the Nile River, AGA is planning to grow 150,000 acres to grow mungbeans to sell to the World Food programme in Pakistan and 200,000 acres mostly with fodder for export. "We soon want to cultivate over million irrigated acres in Sudan", the farm manager firmly stated. AGA reiterates the aspirations to mobilize Sudanese resources to feed the world on its website: "Sudan is a massive country blessed with abundant natural resources and resourceful people, moulded by a harsh physical environment and challenging trading conditions. AGA Group has successfully built a diverse and resilient organization, with a focus on key capabilities, in order to not only survive, but to thrive, in these unique conditions. In many cases, our capabilities not only lead the industries in Sudan, but in the broader region as well." (AGA GROUP 2018).

*Figure 5.3: Rotating pivot irrigation system supplying alfalfa plants at AGA's Waha farm*

For AGA, the Waha farm is not a potential for development but development in the making. In its account of Waha it points to its technological capacity as central to its success. When discussing the high yields of Waha, the manager gives the example that *"if you want to spread this half litre bottle of fertilizer over exactly one hectare, none of the Gezira farmers can do it."* Presenting itself as possessing the kind of advanced technical expertise needed to boost crop productivity conveniently draws the attention away from how AGA is implicated in making of the highly speculative Sudanese food market. It is this involvement that is so despised by Osman and Nour.

The water accounting maps produced by our team are instrumental in rendering AGA as the technical and entrepreneurial leader in a to-be-modernized waterscape. Representing AGA's farm at Waha as a set of highly productive pixels in a sea of unproductivity has been key to attract interest and support from international organisations and investors. Depicting the terrain around the Waha farm as unproductive marks it as a potential target for efforts to 'upscale' the Waha model. Vice versa, AGA's high tech irrigation is well suited to feature as productive on the water accounting map. The perfectly circled geometry of the plots, the accurate measurement of water supplied to each circle and the year round cultivation makes them particularly compatible with efforts to count Nile water benefits highlighted in FAO's 'Information Products for Nile Basin Water Resources Management' (FAO 2011), 'Regional Collaborative Strategy on Sustainable Agricultural Water Management and Food Security' (FAO 2015) and the Nile Basin Initiative's 'Nile Information System' (NBI 2016). Never failing to stress its position outside politics in sparse media appearances (FT 2014, BBC 2013, CNN 2018), AGA works effectively to hide how it became big through the very politics of grain. To understand how AGA's project of high tech agriculture does not proceed smoothly or without limitations, I turn to those rendered as unproductive.

## 5.6  Water accounts as a betrayal of science by politics

Despite the labelling of the plots around the AGA farm as unproductive, and against suggestions by the Government of Sudan (2011), IMF (2014), African Development Bank (2016) and the Government of Sudan (2008) to 'scale up' high yield export agriculture, the size of the AGA's farm has kept its 138 circles. In the story of AGA's manager, the construction of the pivot irrigation farm at Waha was a matter of simply replacing an old inefficient irrigation system. This account is disputed by those who were in charge of production at Waha before AGA took over.

From the early 1990s, Kamal was formally heading agricultural production at the place where AGA now has its farm. As a president representing 1800 tenants who formed the Three Capital Cooperative Society, farm management was very different from today. At his house, which still functions as the office of the Cooperative Society, he explained how a group of agriculturalists, engineers and lawyers founded the Cooperative Society in 1985 with the aim of collectively developing agricultural projects in Sudan. At the time, food prices were rapidly rising and president Numeri had just fallen. Constraints imposed by the IMF led to a sharp reduction in government investment in aging irrigation infrastructure. This put both the experts in the bureaucracy, and the tenancies many of them owned in publicly managed irrigation schemes, increasingly under pressure (Chapter 4).

The idle pumping station of Waha was the first project the Cooperative Society had identified in the late 1980s. The pumps had been built some ten years earlier as part of a 5-7 billion USD portfolio of Arab investments that sought to convert Sudan into the Arab Bread basket and make the Arab world less dependent on the West (Woertz 2013, 171). This was also a way for Gulf countries to reinvest oil money at a time of global financial crisis. Investors from the Gulf had set up – under auspices of the World Bank and the IMF – the Kuwait based Arab Fund which was based on Sudanese land, Arab money and western technology (El-Battahani 1999). By 1977 a Kuwaiti state company had financed the construction of a pumping station that would enable irrigation from the Blue Nile. When the contractor started the construction of irrigation canals the people living around Waha had protested, because their *wadi* for sorghum cultivation and grazing lands would be affected. During protests several were arrested and taken to prison. After this, Kamal tells us, the people from Madhi (see Figure 5.1) and its neighbouring villages destroyed everything except for the guarded pumping station.

For years the pumping station stood idle. When the Cooperative Society was established in the mid 1980s, they identified the 'brand new' pumping station as an excellent opportunity for agricultural development of the Nile bank at Waha. They approached the Kuwaiti owners of the pumps and proposed that the Cooperative Society would talk to the leaders of some of the opposing villages. In exchange for continued access for their animals to graze in the irrigation scheme after harvest and control over 25% of the command area of the irrigation scheme to be constructed, the leaders of four villages along the Nile agreed to cease their opposition to the project. The members of the Cooperative Society, most of whom lived in Khartoum, would lease the remainder of tenancies.

When the military government took over in 1989, it agreed to institutionalize the agreement and lease out the land for the project, but only when the project would help solve one of its pressing problems: To enlarge its grip over the country the military government had forced a large number of military officials into retirement. To provide some of them with a 'pension', 20,000 acres were added to the scheme for 'the Military' and handed out as tenancies to 'retired' military officers. Kamal describes the years that followed in the early 1990s as a grand success. Kuwaiti investors joined the opening of the project in 1990. For three years the Cooperative Society was in business and hoped that it would expand the modest profits it made from preparing the land, distributing inputs and selling cotton, wheat, sesame and some of the vegetables cultivated on the tenancies. With pride Kamal showed an old map of the irrigation grid with rectangular plots that have the exact same dimensions as the plots of the colonial Gezira irrigation scheme discussed in chapter 4 (Figure 5.4).

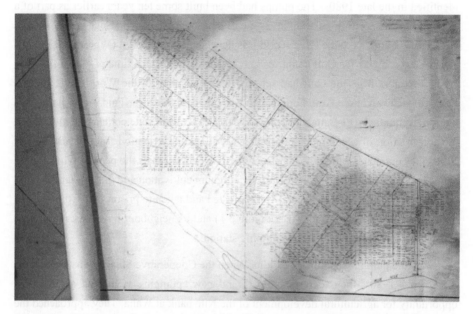

*Figure 5.4: Cooperative Society map to build the Waha irrigation scheme in the late 1980s/early 1990s.*

The 'success' of the Cooperative Society did not last long. After the new Beshir government sided with Iraq in the First Gulf War in 1991, Kuwait stopped investing in Sudan. Prices of inputs went up, spare parts were no longer available and the Kuwaiti company no longer paid running and maintenance cost of the main canal system. The management of the pumping station and canal infrastructure was briefly taken over by Khartoum State and the Cooperative Society shifted to cultivation of local crops such as sorghum and groundnuts. As there was very little reinvestment, production faltered and many of the labourers employed by the scheme left the area.[89] When money started to flow into Sudan again, and land and food prices rose sharply in the early 2000s (Mohammed et al. 2008), the Kuwaiti investors wanted to sell the pumping station. The

---

[89] Eager to get rid of strongholds of bureaucratic power of previous regimes that persisted in the sizable agricultural and irrigation bureaucracies and unions, the government reoriented its attention towards oil production, real estate and communications (El-Battahani and Woodward 2013).

Cooperative Society was intending to collect money for its purchase but soon found out the government was no longer willing to lease the land to the Cooperative Society. Kamal discovered that the government had agreed with the Kuwaiti to instead sell the pump and canals to AGA Group which would become the new leaseholder of the area. Arguing for the need to pursue a new model of the future of Sudanese agriculture (GoS 2008), the government reasserted control over the land it had previously leased to bureaucrats and officers who had become increasingly hostile to government.

For Kamal, the different productivities of the plots of AGA and the remaining plots established by the Cooperative Society cannot be meaningfully explained by simply referring to variations in resilience of individual farmers or differences in access to information. Neither is it useful to him to attribute the higher yields of AGA to an increased efficiency of land and water management. That the state unilaterally liquidated the company managing the farm and stopped fulfilling its maintenance obligations (GoS presidential resolution 324/1995) and put the maintenance burden on the Cooperative Society was unjust. That it subsequently leased the land to AGA he found outright criminal.

For Kamal, the distinct pattern of irrigation that has since emerged (the replacement of some 75% of the smallholder plots by AGA's irrigation pivots) at Waha embodies "not a scientific but a political decision" (Interview with Kamal 5-10-2016). His defence of the model of the Cooperative Society is not very different from AGA. Kamal insists in the superiority of the tenant model of cultivation as rational and scientific. He is convinced that "with proper land and water management" the smallholder plots around AGA's farm, can produce much more than is currently the case. Confronting Kamal with the similarity of his appeal to that of AGA's, he replied with anger: "Who says it is scientific if one farmer displaces more than 1800 others?" After the expropriation of their land, Kamal had not returned. Instead, he sued the government. For him, and most of tenants that formed the Cooperative Society, the Waha irrigation scheme has since become the matter of a court case.

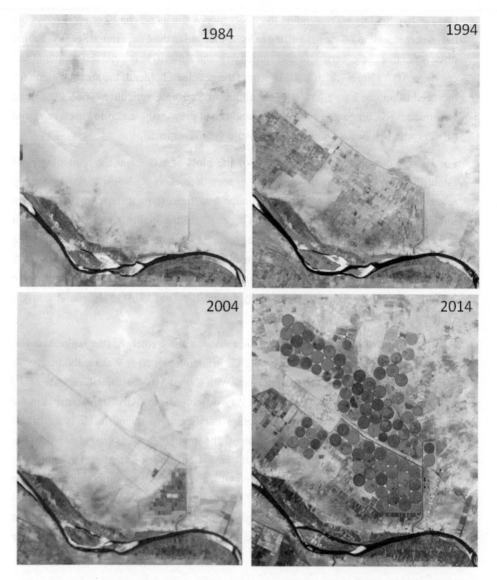

*Figure 5.5: Greenness index for Waha area derived from satellite images made in October 1984, October 1994, October 2004 and October 2014 respectively. The high greenness of irrigated areas with crops shows red on the maps. (insert scale bar area approx 15x 18km)*

## 5.7    Water accounts as distributions of land, work, pensions and food in Waha

Both the accounts of AGA's manager and Kamal make sense when looking at the rectangular and circular plots on the respective satellite images of Waha of early 1994 and 2014 (Figure 5.5). Yet these accounts help little to understand how the rectangles and circles are related. Despite Kamal and Khalid's radically opposing views on 'productive farming', their accounts are similar in their silence about the roles of those who work on the lands of Waha. AGA's manager has little to say about the laborers who guard pumps or weed the crops amidst high tech pivot irrigation installations. While for Kamal his relations to the people of Madhi were key to his brief success, they feature only as a population that needed to be compensated for land that was taken away from them. Representing those who invested in, and worked at, Waha by a single category of productivity, hides how AGA's project to feed the world transformed producers of, and investors in, food to consumers of AGA's imported wheat.

The Cooperative Society regards its members as farmers, but most of them rented out their tenancy or employed a wakil [manager] from Madhi who in turn relied on wage labourers from labour camps and surrounding villages to do most of the actual work. In 1989, a labour camp had been set up near Madhi by 'migrant labourers' who crossed the river from Gezira to work at Waha. More labor was provided by the 'pastoralist' villages of El Hosh which were situated some 5-10 km away from the river. The inhabitants of El Hosh permanently settled and worked seasonally in flood irrigation for those in Madhi since the late 19th century. The predecessors of the 'migrant workers' inhabited the Gezira across the river since the early 20th century. Nevertheless both groups of people are still widely regarded as 'migrants' who were, and still are, not entitled to own or lease the land.

When AGA took over, and heavily mechanized the cultivation, the 'migrant' camp closed. For weeding, both AGA and the remaining owners of agricultural land in Madhi recruited mostly women from El Hosh and neighbouring villages. The overlapping grids of rectangular and circular irrigation thus coincided in consolidating the position of outsiders as owners/leaseholders, people of Madhi as managers, and western migrants and El Hosh as workers. Yet in the process of rearranging irrigation at Waha their positions had changed in an important way: whereas before investors, managers and workers were involved in production to complement their income and food supplies, today almost all at Waha have become dependent on AGA's imported wheat. It was this reconfiguration that shaped both the power of, and the limits to, AGA's success.

During my first week in Madhi, I came to realize that the absence of those who worked the land from accounts of water in and beyond Khartoum was more than just an omission. Despite family connections of the project team which ensured that we were warmly welcomed in Madhi, all but the security agents who visited us on the first evening were reluctant to talk to us about Waha. It was only when Idris was introduced to us a month later that I began to understand more about relations of farming at Waha. Idris was one of the few tenants of the Cooperative Society who had continued to be involved in Madhi. Despite the fact that he lived in Khartoum, some of the village leaders suggested he could tell us about farming in the area.Born and raised in Madhi, he made a career as an agricultural engineer in the Gezira. His position as a professional of the agricultural bureaucracy enabled him – as one of the few people in Madhi – to become a member and obtain (against payment) a tenancy of the Three Capitals Society. The data typed on a A4 with the title 'historical notes of the Waha scheme,' that Idris gave us when meeting with our team, strikingly resembled the information provided by Kamal. It was only when Idris took us to one of the fields still in his possession, when we learned more about the practice of farming at Waha today. Just outside the AGA farm our research team encountered a group of women who worked for the investor that sharecropped Idris's land. Idris and one of these women, Maaza, told us about their involvements in farming. To understand how these involvements had changed over the past 15 years, requires shifting attention to their respective links to another actor: AGA.

Idris stresses how the construction of AGA's farm, like the construction of irrigation infrastructure in the late 1980s, had depended on cooperation with the leaders of Madhi:

> "At first, the people in Madhi did not care, because irrigation stopped. Also they suffered from drought at the time. They thought Ahmed [AGA's CEO - the name by which the pivot irrigation farm is identified around Waha] was just a government body. But when they fenced the land to stop the people and animals from entering we became angry. We protested.  People wanted to increase the (irrigated land) piece of Madhi and hoped that Ahmed would go into an adjacent area.  Security guards treated us badly. They came here on a Thursday and Friday and they took many people to jail for 2 days and they whipped them. Thereafter the canal construction continued and they gave us [the village of Madhi] only less than half than what we used to have, promising more would follow with the second phase of the project."

To compensate for this 'loss' to the company, the government granted it the land that used to be used by the 'pastoralist' village of El Hosh on the other side of the farm. Some of their leaders received a compensation. As one of Maaza's uncles recounts: "This was not the first time this happened. On many sides the companies are taking land around us. Now we are surrounded by companies. Soon we will only be left with our graveyard. Three years ago, with a large group of people we stopped the machines of Ahmed, but they came back with police officers who jailed us".

Despite this history of violence, Maaza and Idris were initially not so negative about the project. The Cooperative Society irrigation scheme had been little productive since 1995. While 80% of the irrigated land in the area had been taken for the construction of the pivot irrigation scheme, AGA's supply of water to the remaining 20% was better and cheaper than it had ever been since 1995. Moreover, the new scheme provided work for Idris and Maaza. Soon after he lost his tenancy, Idris also lost his job in the Gezira scheme, which dismissed the remaining public servants in 2009. In 2010, AGA employed Idris to manage community drinking water projects in Madhi to ensure its continued support. Maaza too was employed by AGA, to work on the weeding of the crops at Waha farm.

Yet by 2015 the growth of the AGA's farm at Waha had come to a halt, and Maaza and Idris's initial enthusiasm was over. With imports of wheat becoming ever more expensive and food prices rising, both AGA's business model and the real wages of its workers have come under strain. Maaza had, like Idris some 15 years earlier, started consuming bread made from imported wheat by AGA. While the produce of the Waha farm is exported as fodder to Saudi Arabia, AGA until 2015 remained the largest of two companies in Sudan to import and mill wheat (FT 2014). And it is here where AGA's biggest challenges were emerging. After the secession of South Sudan in 2011, government income from oil declined by some 75%, and the provision of subsidized fuel and wheat have become under strain. While the fuel price increase led to an increase in the operating costs of diesel powered pumps supplying water to the irrigation pivots, the withdrawal of wheat subsidies took form in the adjustment of the favorable exchange rate through which AGA could import its wheat. When the US dollar exchange rate for wheat imports was adjusted from SDG 2.9 a dollar to SDG 4.0 a dollar in July 2015 (still well below the black market exchange rate of 9 SDG a dollar at the time), AGA halted its flour mill and shortages and bread riots followed soon after (Dabanga 2015). Maaza recalls:

> "At the time, bread prices rose from 3 breads per SDG to 1 bread per SDG and the breads got smaller. We and most in El Hosh refused to buy it. Now the bread price is 2 breads per SDG again but that is still much higher than the 4 to one SDG a year ago and 3 breads to one SDG 6 only months ago."

Over the same year (2014/2015) wages only went up from 25 SDG to 30 SDG. During four months of the year Maaza spends more than two working days wages per week on sorghum and bread, making it very difficult for her to make a living.

Maaza does not blame AGA for the high wheat prices. To her "Ahmed Aziz [AGA's CEO] is a good investor with a lot of money so he can do what he wants". In a way, her explanation seems to underwrite the familiar rendering of investors as free-spirited and ingenious entrepreneurs. Yet to Maaza this ingenuity has little to do with efficiency or technological superiority. She attributes AGA's success to its wealth, which enables it to secure the right ties to land, water, machinery and labor, and maintain its privileged position in Sudanese farming. That these privileges are not self-evident or everlasting became clear when the National Intelligence and Security Service set up its own flourmill in 2014. This reduced AGA's bargaining power in obtaining new concessions of land and water, but did not decrease the price of bread for Maaza.

While on the one hand Maaza and Idris pay the price for AGA's strategizing to hold on to its privileged position in the Sudanese food market, they also shape the boundaries of AGA's success. Over the years both Idris and Maaza stopped working for the company. Idris resigned when the community projects failed to materialize and his position in the village became increasingly difficult. Like many in Madhi, he now blames AGA for its broken promises and the negative effects of the Waha farm on the crops and health of the people in the village. Many, including the village doctor, point at the year round cultivation and intensive use of pesticides by AGA to explain the high prevalence of malaria and cancer in Madhi. Maaza too stopped working for AGA. "Not only is the pay limited. The tight control of working hours and supervision of the work make the work little appealing", she explains. After the El Hosh protest against the expansion, AGA stopped hiring workers from El Hosh altogether. While AGA has little difficulties in drawing new workers from Khartoum and beyond, this further increased tensions between AGA and people from El Hosh. More significantly for AGA, Maaza, and many with her in Sudan found themselves forced to reduce their bread purchases as prices are rapidly rising and bread is not always available (Dabanga 2018, Reuters 2018). While foreign donors still believe in AGA's model of entrepreneurship , AGA finds it increasingly difficult to access wheat for its flour mill and to find internal political support to expand its production for export markets. With 'markets' for import and export ever more uncertain, land and water ever more contested and fuel costs rising, AGA has become increasingly politically and economically isolated within Sudan. The tension between what Mann calls 'expanding aspirations [of the Sudanese people] and private opportunity' (Mann 2013, p300) is undermining both the export oriented business model of AGA and the Sudanese regime.

Since a new round of subsidy cuts in November 2018, the availability of bread and other consumables has been erratic and 'bread riots' have swept the country. In the crackdown of the protests that followed security forces killed at least 50 and detained thousands (AP

2019). In response, more demonstrations were organised around the country to protest the violence used by the National Intelligence and Security Service. One of these protests was staged by employees of AGA who urged the government to end violence and to step down (Dabanga 2019, LUA 2019) (Figure 5.6).

*Figure 5.6: AGA employees protesting violence by security forces in response to bread riots in front of the AGA flour mill, Khartoum 12 February 2019. (Picture: LUA 2019)*

While Sudanese land, water, labor and credit are being increasingly appropriated in the name of 'global agriculture', AGA and people at Waha are not becoming productive or resilient in the ways anticipated by the Government of Sudan, the IMF or the FAO (GoS 2015, IMF 2017, FAO 2016). Mobilizing science, infrastructure and institutions inscribed by earlier relations of land and water use at Waha, the actors featuring in the above accounts shape the water accounting patterns featuring on the maps in ways that no one in our department could have predicted. Instead of expanding the area grown by alfalfa for export, all the overlapping investments in Waha that have created new opportunities appear as unproductive on the maps of Osman, AGA and Kamal.

One of these opportunities has been created by the shifting frontier of appropriation of another substance that makes the Nile: silt. When the Khartoum real estate market thrived after Sudanese oil exports boomed in the mid 2000s, pollution from brick making in the Khartoum area increased dramatically. After brick making was banned in Khartoum, Madhi – 50 km up the river – emerged as a new site for mining clay to supply bricks.

Since 2010 over 300 brick ovens baking Nile clay have emerged along the Nile in Madhi (Figure 5.7). Men from Madhi are once again employed as 'wakils' managers by investors from Khartoum, this time managing brick ovens instead of tenancies. Others leased trucks and set up their own transport business. Both the ovens and trucks draw labourers from the new camp established near Madhi, which was set up by people fleeing from conflict ridden South Sudan. By 2018 the limits to this new frontier of appropriation are already emerging. The bubble of Khartoum real estate is bursting and rising food prices put wages of this business under pressure too. Moreover the supply of key resource for brick making, Nile clay, is being threatened now that the filling of Ethiopia's Grand Ethiopian Renaissance dam has started and the vast majority of Nile sediments will be trapped upstream.

*Figure 5.7: Workers from South Sudan working at a brick making factory near Madhi*

As the limits to the current distribution of land, water, food and silt emerge, Idris, Maaza, and AGA are engaging in new ways of organising the land and water at Waha. As life in Khartoum is becoming prohibitively expensive, Idris considers returning to Madhi, where he plans to expand his investments in farming. Through his ties with the Agricultural ministry, he has managed to register an additional plot in his name that he seeks to connect to AGA's irrigation water. Since last year, he sharecrops with the same investor who employs Maaza. She returned to work for this investor with whom she feels more comfortable. Even though the wages are similar to AGA's, she explains that his personal relation to her "brings [her] a better life". Meanwhile, her brother has invested in growing wheat on the plot which is partly owned by Maaza. AGA, in turn, has succeeded in persuading foreign donors to invest in its plans for turning the company into Sudan's major integrated food company. Supported by a 75 million dollar loan from the African Development Bank, the company stepped up its stakes in the Sudanese food production and processing (AfDB 2018). None of these new ventures is as "productive" as AGA's export farm at Waha, and yet for Idris, Maaza and AGA they provide possibilities that they perceive as more promising.

## 5.8     Conclusion - Accounting for Nile waters

This chapter examined how distinctions between facts and value, and between productive and unproductive forms of water development materialize through projects of water science and development. These distinctions are the historical *products of* practices of 'scientific' representation, privileged concession, and dispossession. At the same time, *they are instrumental to* highly particular projects of the appropriation of land and water. On the East Bank of the Nile near Khartoum, subsequent hydraulic projects (re)produce particular racialized and gendered differences between investors, managers and workers. The mapping analyses in my own academic department at IHE Delft work to reproduce a particular structuring of the roles and disciplines of water science. This chapter highlights how these varied engagements in the making of water development at Waha are much more complex than suggested by the categories of crop and water productivity and how these engagements are instrumental in making these categories anew. To examine what this implies for the limits and possibilities of water accounting tools and culture I return to my own academic department.

During the closing workshop of our 'Accounting for Nile Waters' project at IHE Delft the links between the different accounts of water development highlighted above featured as the central topic. On the first day, students and project staff opened the first version of the Nile Water Lab which brings to the fore the above 'accounts of distribution of land, water, work and food at Waha by Osman, AGA, and Kamal, Maaza and Idris' (Sections 5.4 to 5.6 – see www.nilewaterlab.org/waha). Many of the conference participants appreciated the variety of narratives of water development at Waha. The sentiment amongst most of them was one of sympathy: many found it interesting to highlight and learn about lunch packages, land deals, the contingencies of the Gulf War, rising wheat prices and food riots – but they were not sure how these stories would help to improve the situation. Also, many felt that the presented divisions of land, water and labor were highly particular to the Waha case. This made it difficult to compare such accounts to the other case studies in the project about Ethiopia and Egypt.

The next day, Osman and the remote sensing team presented a preview of the project's final report on "Water Productivity in the Eastern Nile: Case Studies from Sudan, Egypt and Ethiopia" (HRC 2016). The report features Waha as one of the case studies that had been analysed using an existing remote sensing script, one that had been improved for use in this project by members of the project team. The report provided a comparison of the water productivity of the three 'cases' of new irrigation development along the Nile. For the Waha case it had proven undoable to calibrate and validate the rainfed crops surrounding AGA's pivot irrigation farm. The project team had therefore focused on the AGA farm which showed yields of 18 ton of alfalfa per hectare per year and a water productivity of $1.0 - 1.8$ kg/m$^3$. While yields were clearly higher than the surrounding plots (for which no values were calibrated), the high water consumption of AGA's farm made its water productivity still lower than both the FAO Reference values and the values of the cases studied in Egypt and Ethiopia. There was thus still a lot of room for improvement.

During the discussion that followed many reiterated the need to support ways to produce ever more food with ever declining amounts of water. At the same time, there was consensus that the science and politics of irrigation at Waha were deeply intertwined and could not be understood as separate from each other. Yet there was no agreement on how to proceed from there. Remote sensing experts hinted at incorporating social indicators in their framework and thus moving towards 'inclusive water accounting'. This fits well with suggestions by CGIAR and FAO to move from 'crops per drop' to 'SDGs per drop' (CGIAR 2015, FAO 2018). True to the water accounting culture, such a way forward insists on the need to incorporate social and ecosystem services that were not previously valued in 'accounts for water' (CGIAR 2015).

This chapter shows not just that many of the relations through which water security is accounted for by people at Waha cannot be captured by the limited possibilities of colouring a 30 x 30 meter Cartesian grid. More importantly, the chapter suggests that promoting a future of resilience which brings social factors into productivity equations might well be instrumental in further marginalizing already disadvantaged people. It would not be the first time at Waha that renderings of people in terms of material and social deprivation would be used to justify new projects that would further weaken the very 'adaptive capacities' mapped out by the proponents of modern development projects. By making the sedimented histories, through which values and interests are fixed into the river bed disappear from view, water accounting maps could provide new 'scientific' ammunition for projects which suggest that people who are not fit for agriculture in terms of the global market should be prepared to accept financial compensation for their land and water. Perhaps even more disturbing, by masking promising new forms of cultivation and collaboration over Nile waters, the water accounting culture works to disconnect these emerging forms of cultivation from circuits of investment and knowledge production that might be key for their consolidation.

This does not mean that the tools of water accounting should be dismissed altogether. On the contrary, if only for their authority in the water accounting culture, maps are a powerful tool to gain recognition in wide circles of policy making, engineering and finance. Osman and Nour understand this extremely well. They used the tools and roles prescribed by the water accounting culture in the Ministry to mobilize funding and cultivate new ties with consultants, companies like Waha, and institutes like IHE Delft. In this way Osman and Nour have used water accounting for re-asserting a form of public control – however limited – over the management of irrigated production. In the Gezira (see chapter 4) this has enabled new cooperative experiments that respond to the energy and hopes of at least some Sudanese producers and consumers rather than speculative investors and dependencies created in the name of their improvement (HRC 2018). This has not been without considerable expense and hesitation. Not only did real wages at the Ministry in Sudan drop by 50% since the start of the project in 2014, Osman and Nour are continuously in doubt about working in a ministry alongside security agents and thus being part of a repressive security apparatus. Their efforts to reengage the state to secure more promising water futures for Sudanese citizens are always in danger of co-option by the networks of speculating entrepreneurs in Sudan and abroad that they seek to discredit.

As an IHE Delft funded researcher interested in exploring new possibilities of making water security along the Nile, I also see great value in mobilizing "water accounting" tools for more emancipatory projects of water security. Not to increase water productivity per se, or to proof that "water accounting" could be abused to justify violent projects, but as a starting point for raising new questions about the making of distributions of knowledge and water. Once we stop discussing about the correctness and transparency of the values on the maps – i.e. key concerns of the water accounting culture – and shift our attention to the practices of making and using maps and the categories displayed on these maps, new questions about making water security at Waha come into view: e.g. who selected certain categories, for what reasons and to whose benefit? And what do these categories get to mean for people who become labelled as productive and unproductive?

These questions not only led us to inquire into how different realities of accounting for water at Waha exist side by side. It redirected attention to how these realities took specific form in contested categories of progress and irregular irrigation grids of Waha. By so doing, this chapter yields two new understandings about the limits and power of the high tech pivot irrigation model of farming at Waha: 1. at this moment, the upscaling of water productivity around Waha is an illusion: the high tech pivot irrigation farm can only continue to remain the much admired 'example for future farming in Sudan' by sticking to, and fortifying, its present boundaries; 2. new opportunities for increasing water and food security at Waha and in Sudan are opening up around the fortified pivot irrigation farm, but on water accounting maps these merely show up as 'unproductive'.

To contribute to making space for some of these emerging opportunities, and therewith hopefully also to increasing water security for more people in Sudan, it is important to work on changing the water accounting culture. Only when we – academics – acknowledge, appreciate and condemn how values of productivity, individual entrepreneurship, private property and transparency, that are institutionalized by the water accounting culture, are often put to work – through academia – to the benefit of few and the detriment of many, can we change this culture and mobilize its tools for more just futures of water security. This means no longer rendering people as being stuck in underdevelopment, and instead appreciating their changing positions and roles in the distribution of land, water, food and power. Researchers and research funding organisations should make a contribution by making explicit, and taking responsibility for, the ways they make some orderings of water possible and not others.

# 6

# Conclusion

## 6.1    Introduction

Water development projects aiming to increase water security are often imagined by scientists and development organisations as technological and governance innovations that will protect populations facing dwindling and ever more uncertain supplies of water (Sadoff and Grey 2007, GWP 2014b, UN Water 2013). Not only do such views of water security foreground a very particular way of living with water as an economic resource. By not recognising earlier collaborations and struggles over land, water, food and labour that formed the river, the limits to water security are often assumed to be natural and universally knowable.

Taking the morphology of particular infrastructures of the river as a starting point, I have sought to understand modern projects of water security as material, cultural and political formations. This enables an understanding of a) how new projects of water security along the Blue Nile are shaped by the struggles of the past that have shaped the river of today, and b) how these projects create new possibilities for distributing Nile waters.

This concluding chapter presents the answers to the following research questions and specifies the contribution of this dissertation to the emerging literature on water security.

1.  Who and what make the modern limits to water security on the Nile and why so?
2.  How do projects in the name of water security on the Nile transform the river and who benefits?
3.  What does this imply for the use of science for understanding and changing the emerging patterns of Nile water distribution?

## 6.2    (Re)making the limits to modern water security on the Nile

The first contribution of this dissertation is that it shows how the categories, units and limits to modern Nile water security (e.g. water resources, population, water productivity, water prices, water as a dividable and tradable property), like Linnaeus' taxonomy of the 18th century, are not natural or given but have been produced with the practices that shaped the organisation of Nile waters during the hydraulic mission era. By analysing how practices of mapping, standardization and calculation took form in projects of modern Nile river development, the dissertation brings into view the intersecting relations of water distribution, investment and water science through which politics of water security are obscured. Building on Mitchell's analysis of the techno-politics of modern Egyptian development (2002), it shows that modern engineers not only constructed the Nile as a matter of optimization, integration and adaptation. By declaring their techno-economic calculations of river development as objective facts that are separated from the values of development, their science became instrumental to the violent appropriation of land, water and labour along the river for industrial development.

Chapter 2 demonstrates how three waves of investment in large scale hydraulic infrastructure over the last 200 years have each shifted the definition of, and limits to, water security along the river. The first wave of construction of large barrages and dams – starting in the 19th century to enable the expansion of cotton cultivation – led to the depletion of almost the entire base flow of the Nile. The concerns that arose in regards to safeguarding this base flow from upstream consumption invoked the invention and conquest of the *Nile Basin* territory. A second wave of hydraulic construction took place from the 1950s to the 1970s. The postcolonial dams that were constructed on the upstream national boundaries of Egypt and Sudan enabled the storage of the peak flows of the Nile in the name of *national* development. With this, the definition of water security and the consumption of Nile waters came to encompass (almost) the entire river flow. Yet this did not stop governments and companies from investing in the contemporary, third, wave of mega-projects on the Nile, which started at the end of the 20th century. Faced with depleted rivers and old irrigation schemes that are no longer producing export commodities, international organisations such as the World Bank, the IMF, FAO and IWMI worked with governments and investors along the Nile to redefine water security in terms of "ecosystems [which] are seen as capital assets, with the potential to generate a stream of vital life-support services meriting careful evaluation and investment" (Turner and Daily 2008, p25).

Chapters 3, 4 and 5 show how the new liberal ideal of water security thus created is mobilized by governments and investors today, to plea for new rearrangements of land and water in a) the Ethiopian highlands where state-led soil conservation programmes arguably constitute the Nile's largest contemporary hydraulic development project in terms of labour, b) irrigation reforms in the Nile's largest irrigation scheme in the Gezira, and c) a modern private centre-pivot irrigation scheme which was built on top of a scheme that used to be collectively managed and is now promoted as Sudan's hope for productive agriculture.

On a hill slope of the Choke Mountains of Ethiopia a new programme of 'integrated participatory watershed development' was set up to protect plots from soil erosion under increasing intensification of cultivation. To implement this programme, government officials set up a 'development army' by mobilizing and supervising individual farmers to construct communal terraces. During implementation, the ditches from which soil was taken to construct the 'terraces' were turned into drains to prevent scouring flows during extreme rain storms. By leading the drainage flows along the 'terraces' to plots that are sharecropped in by young landless families, new inter-generational rivalries are created between these families and older land holding families. Party officials take sides in the conflict by recognizing the old landholding generation as 'true farmers' concerned with their soils and therefore leaders of the 'development army'. By directing control over training in 'agriculture and development', dispute resolution and input distribution to the elderly landholding generation, they create a strong support base for the ruling party. The 'participatory' configuration of drains thus sustains the ever-deepening social and physical cracks in the hill.

Some 700 km downstream along the Blue Nile, in the century old Gezira irrigation scheme a new Irrigation Management Transfer policy that was designed by the government and the World Bank to increase the scheme's water efficiency and economic performance is being implemented. An analysis of the flows of sediment, water and crops along one of the scheme's 1,500 canals shows how a powerful network of male tenant investors has taken control over the newly established water user associations and with it the scheme's agricultural service provision and canal maintenance. Their concern with the amount of cubic meters silt excavated rather than the shape of the canal has resulted in increased differences in water access along the canal. On lands with good access to water, that are largely under control of rich investors, male migrant sharecroppers and labourers are hired to grow high value cotton and vegetables. In contrast, the majority of lands that have increasingly unreliable access to water are sharecropped by migrant women from poorer tenants who have largely withdrawn from agriculture altogether. These women – numbering 100,000s in the Gezira Scheme but hardly featuring in official reports – deal with large uncertainties in access to water by cultivating drought resistant staple crops and prioritising work plots where water is available. Their low yields per hectare are abused by the government to demand for yet more radical reforms and explain its withdrawal from the irrigation scheme altogether.

The other bank of the river shows what such 'radical reforms' look like in an irrigation scheme that has been partly demolished and is being overwritten by a high tech centre-pivot irrigation scheme that is presented as the example of future productive agriculture in Sudan and beyond. Being involved in a CGIAR funded project seeking to foreground and discuss different realities of water productivity, we found how the 'water productivity maps' that gain popularity in embassies, FAO and IMF, fit particularly well with highly mechanised farming. What these 'water productivity maps' do not show, is how the high productivities of the new farm were made possible by land deals between a regime struggling to import wheat to feed the urban poor and a company that until recently controlled more than 50% of those imports. While mobilizing high yields that are also visualized by the satellite assessment to argue for an expansion of its highly mechanized farming model, the company remains silent on fierce opposition from surrounding farmers who associate the farm with corrupted land deals, increased incidence of pests and even cancer.

All three projects of water security socialize people as individual asset owning entrepreneurs seeking to generate new streams of benefits by increasing *irrigation efficiencies* and *adapting to rising uncertainties*. Failure to achieve the intended objectives of 'efficiency and resilience' is mobilized to demand for 'upscaling' efforts and freeing up more land and water for reform. In doing so, the projects normalize particular class, gender, racial and generational relations that shape and consolidate highly uneven patterns of access to water.

## 6.3    Rearranging patterns and possibilities of accessing and knowing Nile water

To point out how the ontologies of modern Nile water security are shifting with the unfolding of the hydraulic mission is one thing; but to account for the irregularities in the patterns of Nile water distribution is another. By analysing the engagement of specific water users, infrastructures and scientists in remaking the river morphology, this dissertation shows how their respective involvements are not confined by the roles assigned to them by modern engineers, planners and visionaries of new water development megaprojects. New projects of water security do not simply 'free up' heavily committed water for a new round of development; as these projects are implemented they are confronted with historical meanings, infrastructures and peoples inscribed into the course of the river during earlier rounds of investment.

This brings me to the second contribution of this dissertation, which concerns the materiality of water security. Theories of water security often have a rather narrow sense of materiality, which treats water as a resource that can be unproblematically substituted and optimized. This obscures the uneven and contested historical relations of work that shape the river and that have created different positions in access to Nile water. While 'political ecologists' and 'world sociologists' have been attentive to the social and historical relations which shaped the distribution of resources around the world, their analysis of how these distributions materialize in the landscape has been limited. This dissertation brings in this material dimension by paying attention to the ways in which the making of infrastructures, institutions and knowledge shapes water security along the Nile. By analysing the river bed as the product of work, the dissertation brings to the fore how the outcomes of material and discursive struggles to secure Nile waters solidified in the river at one point in time shape specific material and scientific positions that come to matter in subsequent struggles over the river. This brings into view a) how multiple meanings, loyalties and material differences that were fixed into the river by large hydraulic developments throw long shadows over the future use of Nile water; b) how the varied agencies which take form through Nile development make the river both increasingly contested and full of possibilities; and c) that investments in and science of modern projects of Nile development have become increasingly speculative and contested.

As shown in Chapter 2, the exponential rise in the production of cotton, sugar, alfalfa, and electricity along the Nile was mortgaged on growing inequalities in welfare, new racialized and gendered divisions of work, and staggering rises in public debt. As capital intensive production for export concentrated large tracts of land along the Nile in the hands of a few, a large share of the Egyptian and Sudanese people became dependent on increasingly expensive wheat imports from the USA, Australia, Russia and Kazakhstan. With land and water increasingly committed to large scale agriculture the outflow of the Nile has been reduced by some 85% with the remainder unsuitable for consumption but required to halt further intrusion of salts in the Delta. Another 'premium' of modern water security is that 25% of Nile waters are now evaporating from man-made reservoirs along the river.

Yet the securitization of the river is not just a story of limitless exploitation and domination. While a new wave of projects in the name of water security is in full swing, the limits to the hydraulic mission are increasingly evident. Not only have increasing landlessness and rising food prices sparked widespread popular resistance in Egypt, Ethiopia and Sudan, new large investments in water development have become more speculative: this is expressed both in the increasing financing of projects with debt and the many lands purchased by developers remaining undeveloped. As shown in Chapters 3, 4 and 5, people that have been portrayed as backward, underdeveloped and lacking resilience by modern projects of river basin development are managing to mobilize old hydraulic infrastructures and institutions for ventures of water security that are hardly compatible with the neoliberal logics of water security that are now popular amongst development organisations and universities around the world.

Chapter 3 demonstrated the limits of "integrated participatory social conservation programmes" in the Choke Mountains. The seemingly powerful 'development army' led by a new coalition of landholders is crumbling as gullies created over new drainage routes are now widening into the lands of powerful landholders. A generation of young landless families increasingly refuses to contribute labour to a technical development programme that seeks to conserve soils on which it no longer depends. Failing to live up to the image of the 'farmer interested in soil conservation', they have created a competing 'trader model' with its own institutions of finance, oxen sharing and barter with lowland markets.

Chapter 4 showed how the migrant women who are structurally ignored by the formal irrigation institutions manage to divert most of the water along a canal in an irrigation scheme that – despite all odds – has attracted an increasing amount of water: in 2011 more than 10% of all Nile water. Working to cultivate sorghum and groundnuts wherever they can access water, they have made possible that the scheme produces more food than ever before, providing grain, vegetables and vegetable oil to an estimated 4 million people.

Chapter 5 showed how efforts to 'upscale' the success of a high-tech private pivot irrigation scheme, which produces alfalfa for export to Saudi Arabia, are not unfolding in ways anticipated by the company which farms the scheme, the Government of Sudan or the IMF. An analysis of changing relations of agriculture and food distribution shows how the recent removal of food subsidies has not – as market believers expected – contributed to further expansion of the high-tech private farm. On the contrary, it has undermined both the 'wheat import – alfalfa export' model that made the farm owner Sudan's largest agri-food corporation. Despite the fact that a pump with the capacity to enlarge the irrigated area is already in place, the upscaling of the 'success' of the model farm remains an illusion. Faced with rising opposition from its neighbours who suffered from earlier expansions and rising food prices, the boundaries of the model farm have been fenced and consolidated. Around the farm new niches of sorghum, vegetable and wheat farming are emerging. Interestingly, some of these initiatives rely on a steady supply of water provided by the company through the tail ends of its canals at water rates that are amongst the lowest in Sudanese irrigation. The investment, water consumption and production of the sorghum, wheat and vegetables for local markets are between 5 and 20 times lower than the high-tech 'model' farm. With low yields, these new initiatives appear on remotely sensed water accounting maps as 'unproductive'. Moreover, the emerging forms of work reproduce highly unequal gender and racial relations. Despite these drawbacks, both local investors and seemingly marginal workers are convinced of the possibilities of the new 'unproductive' modes of investment emerging on the fringes of the corporate farm. The returns to investment of producing food for local markets and conditions of work offered by the investors from neighbouring villages with whom they fought against the company, make them committed to work on ventures that few outside Waha recognize as promising.

Together, the chapters situate the engagements of water users, infrastructure and investors of three Nile development projects in a) flows of water and sediment, b) narratives of environmental transformation, and c) relations of investment in agriculture and energy production that connect the projects. The 'incorporated comparison' of the Nile development project thus created shows: (1) how these projects contribute to powerful networks of agro-corporations (like AGA), development banks (like the World Bank, agricultural research funds (like CGIAR) and education institutes (like TU Delft and IHE Delft), and government officials (like those promoting Aid for Trade) that pursue the expansion of corporate control over land, water, labour and capital in the name of environmental accounting, participation and resilience; and (2) how the productive returns to the financial and scientific investments of these networks are increasingly limited. Problems to draw labour, water and private investment for new mega-projects of hydraulic development along the Nile make it ever more obvious that populations do not become productive in ways imagined in neoliberal policies of water security. With land, water and labour increasingly appropriated by modern hydraulic projects of development, financial investments in modern Nile development have become more speculative. Rather than thriving productive investments in export agriculture, we see loss making sugar irrigation schemes (Chapter 2, see also Fantini et al. 2017), land leases merely functioning as hedges against further rises in the food markets, and aspirations for scaling up of large scale production of irrigated alfalfa for export to Saudi Arabia faltering (Chapter 5). With opportunities for 'saving' water by increasing irrigation efficiencies being very limited (Chapter 2, see also Molle et al. 2018) and the environmental and political costs of diverting water to new hypermodern schemes rising (Chapter 2), the prospects for new megaprojects are reducing. The targeted subjects of modern river development make use of the new spaces for 'river development' thus created by carving out their own projects. They creatively mobilize old irrigation and drainage infrastructures in ways that escape the universal logic of modern water security (Chapters 3, 4, 5).

## 6.4    Water security beyond modern theory

Why does the focus on the recursive relations between the making of science and river matter? Who should act upon it? And what is the role of this dissertation in these relations?

As highlighted in Chapter 5, today's culture of water science is increasingly focused on making water flows transparent for entrepreneurs in order to optimize their production and adapt to increasing financial and climate uncertainties. In the process, accountability is reduced to narrow terms of accounting: market transactions demand the quantification of water, crops and infrastructural coverage in ways that can be expressed and exchanged as ecosystem services serving the global population (e.g. CGIAR 2014, UN Water 2016). With biomass production, water productivity and Nile populations reduced to gridded entities readily available on maps with a click of the mouse (NBI 2016, FAO 2018), it becomes easier to perceive them as abstract and alienated from their socio-ecological relations through which they are positioned vis-à-vis each other. This is deeply problematic for the ways in which it reshapes both academia and the Nile River. By making the sedimented histories through which values and interests are fixed into the river bed disappear from view, the emerging scientific culture provides new 'scientific' ammunition for neo-colonial projects which suggest that people who are not fit for agriculture in terms of the global market should be prepared to accept financial compensation for their land and water. As in the previous waves of hydraulic investment in the Nile, this hides how new projects of hydraulic development create the very externalities – more people without land, further deterioration of water quality, rising public debt and prices of food – that fuel the extension of the cornucopian dream of 'green growth'. Equally disturbing, it frustrates promising new forms of cultivation and collaboration over Nile waters from making wider connections of science and investment that are crucial for their consolidation.

The morphological account of Nile development offered in this dissertation seeks to engage with and change this water accounting culture by analysing a) how facts and values of water security intertwine as they solidify in the river bed, and b) how the identities and opportunities of investors in water development take form with the river. Here I am inspired by Goethe's idea that no ordering of the environment can be separated from the ends of that ordering (Goethe 1987 [1776-1832]) and contemporary accounts that seek to understand globalization beyond singular terms without lapsing into relativism (Harding 2015, Haraway 2016, Moore 2015). These accounts do not only provide the tools to rethink conventional categories of water security (water resources, population, water productivity, individual shares of water) and understand emerging patterns of collaboration over Nile waters. The insistence on the intricate connections between the metrics and materiality of development takes us to exciting debates about challenges of 'science for development'.

Globalizing trade and rapid urbanisation pose new challenges to universities and bureaucracies divided by disciplinary and sectoral boundaries (Beck 1992). As the roles and rules of experts of development are increasingly openly questioned by the people they long claimed to work for (Mitchell 2002, Esteva and Escobar 2017), water experts are pushed by governments and development organisations to worry about the impact of their scientific studies, policies and technological innovations (e.g. Figueres et al. 2017). As technologies and metrics for development are reassembled in the name of green growth, old racisms, sexisms and class differences take new directions and forms.

Projects promoting 'citizen science', 'data democracy' and 'benefit sharing' for sustainable development, however well-intentioned by many of their protagonists, too often remain the brainchildren of white male professors from Europe and the US drawing up theoretical frames for case studies to be implemented by Southern scholars and filled by lay 'people' (cf. Wekker 2016). New models of 'aid for trade' condition the granting of public funds for research in 'the Global South' on alignment with corporate interests, forcing NGOs to increasingly adopt values of corporate enterprise (Savelli et al. 2018).

This dissertation draws from the tools of 'metamorphosis' (Goethe 1987), 'situated knowledge' (Haraway 1991), 'incorporated comparison' (McMichael 1992) and 'world ecology' (Moore 2015) – all double edged concepts to BOTH bring into view the politics of categorization AND recover responsibilities for the material/ecological implications of their use – to analyse how the making of 'science for development' in the age of market environmentalism reconfigures relations between the Nile river, the drawers of its water and scientists like myself who study it.

The morphological approach to river development presented in this dissertation enabled a relational analysis of a) how the superimposition of colonial infrastructures and discourses of development justified violent appropriations of land, water and labour, and b) the varied agencies of water users, infrastructures and scientists in shaping new forms of water distribution and cultivation. It traced how intersections of the changing techno-scientific apparatus of water security, investments in agriculture and distribution of water and sediment, created and located 'specific places' (Massey 1993, p66) of Nile river basin development. In this way, the dissertation simultaneously brings into view how categories for knowing river development change with the river and it mobilizes the same categories to keep track of material redistributions of wealth this enabled.

This juggling to capture the relations between the contingent practices that make specific Nile places and the abstractions from these places required to recover accountabilities for redistributions of projects of modern Nile development yields a morphological account of river basin development that is inherently troubled[90]. In many ways my account is not so different from accounts of many of the colonial engineers of the late 19th century. To name a few clear reasons: (1) I share with the colonial Nile engineers of the 19th century much of the scientific language and a familiarity with Cartesian depictions that are rooted not only in the Nile River but also in Europe and the scientific revolution. (2) In the contemporary water accounting culture, the use of modern scientific English language

---

[90] For an inspiring account of why these troubles are more than simply drawbacks, see Haraway's 'Staying with the Trouble' (Haraway 2016).

and Cartesian maps is a condition for funding and success. (3) The use of Cartesian maps and academic English enables discussions in classrooms and international organisations. And last but not least, (4) the use of familiar formats of questionnaires, balance sheet tables and Cartesian maps remain a precondition for obtaining research permissions and a necessary cover for guaranteeing the safety of my co-researchers.

Despite the many convoluted sentences in an attempt to bring in recursivity and relationality to this dissertation, the losses in translating the experiences and stories that informed this dissertation remain considerable. Many of these losses are both inherent to the translation process and necessary to discuss and change the water accounting culture. Yet, it is in reducing some of these losses – e.g. by investing in new ways of researcher engagement and activism – in which I see the potential for taking the morphological approach much further.

Here, I think especially of possibilities of more activist engagements in changing practices of hydraulic development. The observations and conversations to learn about practices of drainage, irrigation and trade that informed this research yielded exciting understandings of emerging limits to the developmental state, participatory irrigation management, and private investment in high value export agriculture (Chapters 3-5). To be sure, we discussed and shared our findings at local and international workshops and we worked to make these part of new curricula for water management in Ethiopia and the Netherlands. Yet, this only created – notwithstanding overwhelming support and close friendships – a severely limited sense of connectivity with those who taught us most about technologies and politics of Nile development. As we deliberately chose to refrain from an explicit activist agenda in the project areas and often treated the people we talked to as informants rather than as companions in furthering overlapping agendas for change, the creative space for understanding and engaging with the irregular patterns of water distribution we encountered was much restricted.

Given this constrained political and research context, I see the greatest promises and challenges for 'science for Nile development' in supporting engaged (and risky) commitments to explore and promote new forms of solidarity that are emerging in the spaces where profits of modern projects of water security are declining. This is very much an ongoing political and material struggle to break down modern separations between science and activism, and between narratives of water security and soil particles that shaped the river. With that, the morphology of the river bed and the morphology of Nile water security become inseparable. Only when science of water security is grounded in material relations through which the river takes form, can its scientists become accountable to and effective in working with those who do not fit modern frameworks of water security. Only then, might we become more productive in contributing to water security for more people along the river.

# References

Abbink, J. (2017). *A Decade of Ethiopia: Politics, Economy and Society 2004–2016*. Leiden: Brill. 253 pp.

Abd Elkreem, T.M.A. (2018). *Power Relations of Development, The Case of Dam Construction in the Nubian Homeland*, Sudan, Reihe: Beiträge zur Afrikaforschung

Abdelkarim, A. (1992) *Primitive Capital Accumulation in the Sudan*, London: Frank Cass

Abdelhadi, A.W. & S. Adam, H & A. Hassan, Mohamed & Hata, Takeshi. (2004). Participatory management: Would it be a turning point in the history of the Gezira scheme?. *Irrigation and Drainage*. 53. 429 - 436. 10.1002/ird.139.

Abdul Salam (2009). *Abdul Salam Committee report on the liquidation and sale of Gezira Scheme assets*, Khartoum, In Arabic

Abiy, A. (2018) PM Abiy Ahmed statement in parliament during presentation of national budget. Available on youtube through: https://newbusinessethiopia.com/economy/ethiopia-approves-12-8-billion-national-budget/, accessed on 6 May 2019

Abo El-Enein, R., M. Sherif, Mohammed Karrou, Theib Oweis, Bogachan Benli, Manzoor Qadir, Hamid Farahani. (2011). Improved water and land productivities in the saline areas of the Nile Delta, in "*Water Benchmarks of CWANA- Improving Water and Land Productivities Improving Water and Land Productivities in Irrigated Systems - Number 10*". Aleppo, Syria: International Center for Agricultural Research in the Dry Areas (ICARDA).

Abu-Zeid, M. (2001). Water pricing in irrigated agriculture, *International Journal of Water Resources and Development*, vol 17, no 4 pp527-538

Acevedo Guerrero, T. (2018). Water infrastructure: A terrain for studying nonhuman agency, power relations, and socioeconomic change. *WIREs Water*, Early Online Version. e1298.

ADB (Asian Development Bank) (2003). *Water for all: the water policy of the Asian Development Bank*

African Development Bank (AfDB) (2016) *Private sector-led economic diversification and development in Sudan*, East Africa Regional Development & Business Delivery Office, Nairobi

African Rights (1997) *Food and Power in Sudan*. London: African Rights.

Agrawal, A. (2005). *Environmentality: Technologies of government and the making of subjects*. Durham, NC: Duke University Press.

Ahlers, R. (2005) *Fixing water to increase its mobility: the neoliberal transformation of a Mexican irrigation district*. PhD dissertation, Cornell University: Cornell.

Ahlers, R., Brandimarte, L., Kleemans, I., Sadat, S.H. (2014) Ambitious development on fragile foundations: Criticalities of current large dam construction in Afghanistan, *Geoforum* 54, 49–58

Ahlers, R., M. Zwarteveen and K. Bakker (2017) Large Dam Development: From Trojan Horse to Pandora's Box. In: B. Flyvbjerg (ed.), *The Oxford Handbook of Megaproject Management*. DOI: 10.1093/oxfordhb/9780198732242.013.27

Ahmed, A.F. (2012) What we know and what we don't know about the Nile - presentation of the Ministry of Water Resources and Irrigation, Cairo 28 January 2012

Akhter, M., & Ormerod, K. J. (2015). The irrigation technozone: State power, expertise, and agrarian development in the U.S. West and British Punjab, 1880–1920, *Geoforum*, 60, 123–132

Ali, T.M.A. (1989) *The Cultivation of Hunger- State and Agriculture in Sudan*, Khartoum, Khartoum University press

Allan, J. A., Keulertz, M., Sojamo, S. & Warner, J. eds. (2013). *Handbook of Land and Water Grabs in Africa: Foreign Direct Investment and Food and Water Security*. Routledge International Handbook. Routledge, Abingdon, UK.

Allan, J.A. (1994) The Nile Basin: water management strategies. In: P.P. Howell and J.A. Allan, eds. *The Nile: Sharing a Scarce Resource - A Historical and Technical Review of Water Management and of Economical and Legal Issues*. Cambridge: Cambridge University Press, 1994, 313-320.

Allan, J.A. (2001) *The Middle East Water Question: Hydropolitics and the Global Economy*. London: I. B. Tauris

Andersson, E., Brogaard, S., Obsson, L. (2011). The political ecology of land degradation. *Annual Review of Environment and Resources* 36: 295–319.

AP (2019) Associated Press Cairo February 8, 2019, Sudan doctors' union says 57 killed in recent protests, https://www.apnews.com/5775125fac91424594f74fc099b9a642, Accessed 5 March 2019

APRP (2000) - Availability and Quality of Agricultural Data for the New Lands in Egypt, Abt. Associates, Cairo (Agricultural policy reform program - monitoring, verification and evaluation unit)

Arab Republic of Egypt, The Federal Democratic Republic of Ethiopia And The Republic of the Sudan (2015) Agreement on Declaration of Principles between The Arab Republic of Egypt, The Federal Democratic Republic of Ethiopia And The Republic of the Sudan on the Grand Ethiopian Renaissance Dam Project (GERDP), Signed in Khartoum on 23 March 2015

Arab Weekly (2017) Egypt mulls its options after failure of Nile dam talks Sunday 03/12/2017, https://thearabweekly.com/egypt-mulls-its-options-after-failure-nile-dam-talks, accessed 5-6-2018

ARDC (2009). Sustainable Agricultural Development Strategy Towards 2030.Ministry of Agriculture and Land Reclamation, Cairo, Egypt (Agricultural Research and Development Council).

ARE (2001) Matching irrigation demands and supplies, APRP report 45

Arrighi, G. (1994). *The Long Twentieth Century*. London: Verso.

Arsano, Y. (2007) *Ethiopia and the Nile: Dilemma of National and Regional Hydro-politics*. Thesis (PhD). Center for Security Studies, Swiss Federal Institute of Technology.

Awadallah, A.G. (2014) Evolution of the Nile River drought risk based on the streamflow record at Aswan station, Egypt, *Civil Engineering and Environmental Systems*, 31:3, 260-269,

Awulachew SB, Yilma AD, Loulseged M, Loiskandl W, Ayana M, Alamirew T. (2007) *Water resources and irrigation development in Ethiopia*. Working Paper 123, pp. 1–3. Colombo, Sri Lanka: International Water Management Institute. (http://www.iwmi.cgiar.org/Publications/Working_Papers/working/WP123.pdf)

Awulachew, S. B.; Merrey, D. J.; Kamara, A. B.; Van Koppen, B.; Penning de Vries, F.; Boelee, E.; Makombe, G. (2005) Experiences and opportunities for promoting small–scale/micro irrigation and rainwater harvesting for food security in Ethiopia. Colombo, Sri Lanka: IWMI. v. 86p. (Working paper 98)

Awulachew, S., Rebelo, L., Molden, D. (2010) The Nile Basin: tapping the unmet agricultural potential of Nile waters, *Water International* Vol. 35 , Iss. 5

Bachewe and Headey (2017) Urban Wage Behaviour and Food Price Inflation in Ethiopia, *The Journal of Development Studies*, Vol.53(8), p.1207-1222

Bakker, K. (2003) *An Uncooperative Commodity. Privatizing Water in England and Wales*. Oxford Geographical and Environmental Studies

Bakker, K. (2010). Privatizing water: Governance failure and the world's urban water crisis. Ithaca, NY, US: Cornell University Press.

Bakker, K., Bridge, G. (2006). Material worlds? Resource geographies and the "matter of nature". *Progress in Human Geography* 30(10): 5–27.

Bakker, K.J., & Morinville, C. (2013). The governance dimensions of water security: a review. Philosophical transactions. Series A, Mathematical, physical, and engineering sciences, 371 2002, 20130116 .

Barnes J. (2012) Pumping possibility: Agricultural expansion through desert reclamation in Egypt. *Social Studies of Science* 42(4): 517–538.

Barnes, J. (2014) Cultivating the Nile: The Everyday Politics of Water in Egypt. Duke University Press, Durham.

Barnes, J. (2014). Mixing waters: The reuse of agricultural drainage water in Egypt. *Geoforum*. 57, 181-191. http://dx.doi.org/10.1016/j.geoforum.2012.11.019.

Barnes, J. and Alatout, S. (2012) Water worlds: Introduction to the Special issue of Social Studies of Science, *Social Studies of Science* 42(4):483-488, DOI: 10.2307/41721338

Barnett, T. (1977) *The Gezira Scheme - An illusion of Development*, London, Frank Cass.

Barnett, T and Abdelkarim, A. (eds.) (1988) Sudan: State, Capital and Transformation, Croom Helm, Beckenham, Kent.

Barnett, T. and Abdelkarim A. (1991) *Sudan: The Gezira Scheme and Agricultural Transition*, London, Frank Cass

Batubara, B., M. Kooy and M. Zwarteveen. (2018). Uneven urbanization through flood infrastructure interventions in (Post-) New Order Jakarta: socio-spatial differentiation and equalization in and beyond the mega-city. *Antipode* (50) 5: 1186-1205. https://doi.org/10.1111/anti.12401

Bayabil, H. K., Tilahun, S. A., Collick, A. S., Yitaferu, B., and Steenhuis, T. S. (2010). Are runoff processes ecologically or topographically driven in the (sub) humid Ethiopian highlands? The case of the Maybar watershed. *Ecohydrology* 3, 457–466. doi: 10.1002/eco.170

BBC (2013) Sudan hopes technology will transform farming 6 December 2013, https://www.bbc.com/news/business-25230126

Beinin, J. (2001) *Workers and Peasants in the Modern Middle East* (Cambridge: CambridgeUniversity Press,

Beck, U. (1992) *Risk Society: Towards a New Modernity*. New Delhi: Sage.

Bhabha, H.K., (1994) *The Location of Culture*; 285 pp. New York: Routledge

Bhrane, A. (2012). Expertise, positions and negotiations about soil and water conservation: Case study of the government SWC program in the Choke Mountains of Ethiopia. Delft, MSc thesis UNESCO-IHE.

Blackmore, D. & Whittington, D. (2009). Opportunities for Cooperative Water Resources Development on the Eastern Nile:Risks and Rewards. An Independent Report of the Scoping Study Team to the Eastern Nile Council of Ministers. WorldBank, Washington, DC, 102 pp.

Blaikie, P.M. (1985). *The political economy of soil erosion in developing countries*. London: Longman.

Blaikie, P.M., Brookfield, H. (1987). *Land degradation and society*. London: Methuen.

Block, P., Strzepek, K. and Rajagopalan, B. (2007) Integrated management of the Blue Nile Basin in Ethiopia: Hydropower and irrigation modeling. IFPRI Discussion Paper 700. Washington D.C.: International Food Policy Research Institute.

Bloomberg (2015) Ethiopia May Ship Sugar in 2016 as India-Backed Plant Ready, by William Davison http://www.bloomberg.com/news/articles/2015-11-13/ethiopia-may-ship-sugar-in-2016-as-india-backed-plant-completed last accessed 6 March 2016

Boelens, R. (2009) The politics of disciplining water rights, Development and Change, 40, 307–331

Bolding, A. (2004) In Hot Water: A Study on Socio-Technical Intervention Models and Practices of Water Use In Smallholder Agriculture, Nyanyadzi, Catchment, Zimbabwe. PhD dissertation, Wageningen, Wageningen University.

Breisinger, C., Olivier, E., Perrihan, A., & Yu, B. (2012). Beyond the Arab awakening: Policies and investments for poverty reduction and food security. Washington DC: International Food Policy Research Institute.

Brown, P., El Gohary, F., Tawfic, M., Hamdy, E., Abdel-Gawad, S., (2003) Nile River Water Quality Management Study, Egypt Water Policy Reform Project, Report No. 67. Ministry of Water Resources and Irrigation and USAID.

Brown, R. (1988) On the Rationale and Effects of the IMF Stabilisation Programme in Sudan under Nimeiry: 1978 to the April 1985 Popular Uprising, Geoforum 19(1) 71-91

Brown, R. (1988) On the Rationale and Effects of the IMF Stabilisation Programme in Sudan under Nimeiry: 1978 to the April 1985 Popular Uprising, Geoforum 19(1) 71-91

Brown, R.H. (1896) History of the Barrage at the head of the Delta of Egypt. Cairo: F. Diemer

Bush, R. (2011) Coalitions for Dispossession and Networks of Resistance? Land, Politics and Agrarian Reform in Egypt, British Journal of Middle Eastern Studies, 38:3, 391-405, DOI: 10.1080/13530194.2011.621700

Butler, J. (1990) Gender Trouble: Feminism and the Subversion of Identity, New York: Routledge.

Capacity Building Office (CBO). (2007) 'Question of good governance: training manual prepared for peasants of rural kebeles', internal document. Bahir Dar: Capacity Building Office.

Cascão, A. E. (2009a). Political economy of water resources management and allocation in the Eastern Nile River Basin. PhD thesis: King's College London, University of London.

Cascão, A. E. (2009). Changing power relations in the Nile River Basin: Unilateralism vs. cooperation. Water Alternatives 2(2), 245–268.

Cascão, A. & Nicol, A. (2016a) GERD: new norms of cooperation in the Nile Basin?, Water International, 41:4, 550-573, DOI: 10.1080/02508060.2016.1180763

Cascão, A. E., & Nicol, A. (2016b). Sudan, 'kingmaker' in a new Nile hydropolitics: Negotiating water and hydraulic infrastructure to expand large-scale irrigation. In E. Sandström, A. Jägerskog, & T. Oestigaard (Eds.), Land and hydropolitics in the Nile River Basin: Challenges and new investments. Abingdon: Routledge-Earthscan.

CGIAR (2007) Bioengineering of Crops Could Help Feed the World: Crop Increases of 10-25 Percent Possible, https://hdl.handle.net/10947/198

CGIAR (2011) Changing Agricultural Research in a Changing World A Strategy and Results Framework for the Reformed CGIAR https://cgspace.cgiar.org/bitstream/handle/10947/5224/CGIAR-SRF-March_2011_BROCHURE.pdf?sequence=1

CGIAR (2014) Ecosystem services and resilience framework. Colombo, Sri Lanka: International Water Management Institute (IWMI). CGIAR Research Program on Water, Land and Ecosystems (WLE). 46p. doi: 10.5337/2014.229

CGIAR (2014) WLE Nile: Open Call for Expressions of Interest (EOIs) 19 August 2018

CGIAR (2015) Putting ecosystems into the SDGs, 2014-2015 highlights, CGIAR Research Program on Water, Land and Ecosystems (WLE), International Water Management Institute (IWMI)

CGIAR Research Program on Water, Land and Ecosystems (WLE). 2014. Ecosystem services and resilience framework. Colombo, Sri Lanka: International Water Management Institute (IWMI). CGIAR Research Program on Water, Land and Ecosystems (WLE). 46p. doi: 10.5337/2014.229

Chesworth, P.M. (1994) "History of water use in the Sudan and Egypt." In The Nile: Sharing a Scarce Resource. A Historical and Technical Review of Water Management and of Economic and Legal Issues, P.P. Howell and J.A. Allan, eds., 65-80. Cambridge: Cambridge University Press.

Chinigò, D. (2014). 'Decentralization and agrarian transformation in Ethiopia: extending the power of the federal state', Critical African Studies 6 (1): 40–56.

Chinigò, D. (2015). 'The politics of land registration in Ethiopia: territorialising state power in the rural milieu', *Review of African Political Economy* 42 (144): 174–89.

Churchill, W.S. (1899) The River War: An Historical Account of the Reconquest of the Soudan, volume II, London, Longmans, Green and Co.

Clarkson, A. I. (2005) Courts, Councils and Citizenship: Political Culture in the Gezira Scheme in Condominium Sudan. Doctoral thesis, Durham University.

Clement, F. (2013) From water productivity to water security: a paradigm shift? In Lankford, B.; Bakker, K.; Zeitoun, M.; Conway, D. (Eds.). Water security: principles, perspectives and practices. London, UK: Routledge. pp.148-165.

CNN (2018) The man running one of Sudan's largest private companies, Marketplace Africa video by CNN, available on https://edition.cnn.com/videos/world/2018/02/28/marketplace-africa-osama-daoud-abdellatif-sudan-dal-group-b.cnn,, accessed 5 May 2019

Coase, R. (1960) The Problem of Social Cost, Journal of Law and Economics. 3 (1): 1–44.

Cobb, S. (1993) Empowerment and Mediation: A Narrative Perspective, Negotiation Journal, July, 245 - 261.

Cook, C. and Bakker, K. (2012) Water security: debating an emerging paradigm Glob. Environ. Change 22 94–102

Cookson-Hills, C., (2013) Engineering the Nile: Irrigation and the British Empire in Egypt, 1882-1914, PhD dissertation Queen's University

Cosgrove, D., Daniels, S. (eds.) (1988). *The iconography of landscape: Essays on the symbolic representation, design and use of past environments.* Cambridge: Cambridge University Press.

Coward Jr, E. W. (1986). State and locality in Asian irrigation development: the property factor. Proceedings of an Invited Seminar Series by International School of Agriculture and resource Development.

Cromer (1908) Evelyn Baring - Lord of Cromer- Modern Egypt. London: MacMillan and Co.,

Crow-Miller, B.; Webber, M. and Rogers, S. 2017. The techno-politics of big infrastructure and the Chinese water machine. Water Alternatives 10(2): 233-249

Culwick, G.M. (1955) Social change in the Gezira Scheme Civilizations 5:2 173-181

Dabanga (2014) Sudan: Farmers' fury at Al Bashir's accusations of theft, 3 December 2014 https://www.dabangasudan.org/en/all-news/article/sudan-farmers-fury-at-al-bashir-s-accusations-of-theft

Dabanga (2015) Technical issues, flour crisis resolved': Sudan's bakery union, 25 April 2015, https://www.dabangasudan.org/en/all-news/article/technical-issues-flour-crisis-resolved-sudan-s-bakery-union

Dabanga (2018) Khartoum witnesses new bread price hikes, 17 July 2018, https://www.dabangasudan.org/en/all-news/article/khartoum-witnesses-new-bread-price-hikes

Dabanga (2019) Sudan uprising: Countrywide response to call by Sudanese Professionals Assoc, 13 February 2019, https://www.dabangasudan.org/en/all-news/article/sudan-uprising-countrywide-response-to-call-by-sudanese-professionals-assoc

Davis, D., & Burke, E. (Eds.). (2011). Ecology and history. Environmental imaginaries of the Middle East and North Africa. Athens: Ohio University Press

De Waal, A. (2005) *Famine That Kills*. Oxford: Oxford University Press.

De Waal, A. (2015) The real politics of the Horn of Africa - Money, War and the Business of Power, Cambridge, Polity Press

Demsetz, H. (1967). Toward a Theory of Property Rights, American Economic Review 57:347–59.

Derr, J. (2011). "Drafting a Map of Colonial Egypt: The 1902 Aswan Dam, Historical Imagination, and the Production of Agricultural Geography." In *Environmental Imaginaries of the Middle East and North Africa*, ed. Diana K. Davis and Edmund Burke III, 136-157. Athens: Ohio University Press

Dessalegn Rahmato (2009). *The peasant and the state: Studies in agrarian change in Ethiopia 1950s–2000s*. Addis Ababa: Addis Ababa University Press.

Dessalegn Rahmato (2014) The perils of development from above: land deals in Ethiopia, African Identities, 12:1, 26-44, DOI: 10.1080/14725843.2014.886431

Dessalegn, M. and Merrey, D.J. 2015. Motor pump revolution in Ethiopia: Promises at a Crossroads. Water Alternatives 8(2): 237-257

Di Nunzio, M. (2014). 'Do not cross the red line': The 2010 general elections, dissent, and political mobilization in urban Ethiopia. *African Affairs*; 113 (452): 409-430. doi: 10.1093/afraf/adu029

Digna, R.F, Wil van der Krogt, Yasir A.Mohamed , Pieter van der Zaag, Stefan Uhlenbrook (2016) The Implication Of Upstream Water Development On Downstream River Basin: The Case Of The Blue Nile River Basin, Proceedings Vol. 6 pp. 221-226, 7th Annual Conference for Postgraduate Studies and Scientific Research, Basic Sciences and Engineering Studies - University of Khartoum, Theme: Scientific Research and Innovation for Sustainable Development in Africa, 20-23 February 2016, Friendship Hall, Khartoum, Sudan

Dilnessa, Z. (1971). Land use study: Ten selected farmers in Yebrage Hawariat. Haile Selassie I University, Addis Ababa

Dixon, M (2014) 'The land grab, finance capital, and food regime restructuring: the case of Egypt'. Review of African Political Economy 41(140): 232-248.

Dominguez- Guzmán, C., A. Verzijl, and M. Zwarteveen (2017) Water Footprints and Pozas: Conversations About Practices and Knowledges of Water Efficiency. Water 9:16.

Dooge J.C.I. (1992). The Manning formula in context. in Yen B.C. (editor) : "Channel flow resistance - Centennial of Manning's formula", Littleton, CO (USA) : Water Resources Publications, 136-185. ISBN 0-918334-72-1.

Du Boys, M.P. (1879): Etudes du regime du Rhone et de l'action exercee par les eaux sur un lit a fond de graviers indefiniment affouillable. Annales des Ponts et Chaussees 5 (18), 141– 195.

Duffield, M. (1983) Change among West African Settlers in Northern Sudan. Review of African Political Economy, 26, 45-59.

Duffield, M. (1990) 'From Emergency to Social Security in Sudan: Part I', Disasters 14(3): 188–203.

Ege, S. (2002). Peasant participation in land reform. The Amhara land redistribution of 1975. In: Bahru Zewde and Siegried Pausewang (eds.) *Ethiopia: The challenge of democracy from below* (61–70). Uppsala: Nordiska Afrikainstitutet and Addis Ababa: Forum for Social Studies.

Ege, S. (2015). Land tenure insecurity in post-certification Amhara, Ethiopia. Paper prepared for Workshop on Agrarian Transformations in Ethiopia (Humboldt-Universität zu Berlin and Goethe-Institut), Addis Ababa 29 September – 1 October 2015

Eguavoen, I.; Derib, S.D.; Deneke, T.T.; McCartney, M.; Otto, B.A. and Billa, S.S. 2012. Digging, damming or diverting? Small-scale irrigation in the Blue Nile basin, Ethiopia. Water Alternatives 5(3): 678-699

Ekbladh, D. (2002). ''Mr. TVA''. Grass-root development, David Lilienthal, and the rise and fall of the Tennessee Valley Authority as a symbol for U.S. overseas development, 1933–1973. Diplomatic History 26 (3), 335–374.

Ekers, M and Prudham, S. ,(2018) The Socioecological Fix: Fixed Capital, Metabolism, and Hegemony, Annals of the American Association of Geographers, Vol. 108 , Iss. 1

El Nour, S. (2015) Small farmers and the revolution in Egypt: the forgotten actors, Contemporary Arab Affairs, Vol. 8 No. 2, April–June 2015; (pp. 198-211) DOI: 10.1080/17550912.2015.1016764

El Shakry, O. (2007) Great Social Laboratory: Subjects of Knowledge in Colonial and Postcolonial Egypt. Stanford: Stanford University Press.

El-Amin K.A. (2002) Peasant Organisations, Participation and the State: Two Case Studies from Rural Sudan, Ch 6 in Romdhane and Moyo (Eds.) Peasant Organisations and the Democratisation Process in Africa, Russell Press, Nottingham, Great Britain

El-Battahani, A. (1999) Economic liberalisation and Institutional Reform in Irrigated Agriculture: The Case of the Gezira Scheme in the Sudan, in Ed. Limam, Institutional Reform and Development in the MENA Region, p241-269

El-Battahani, A. and Woodward, P. (2013) The political economy of the Comprehensive Peace Agreement in Sudan, pp277-292 in: (Eds.) BerAGA and Zaum, The Political Economy of State Building, Routledge, Oxen and New York.

El-Zain, M, (2007) Environmental scarcity, hydropolitics, and the Nile. Maastricht, Netherlands: Shaker Publishing

Elnour, F.M. and Elamin, E.M. (2014) Natural Resources, Agricultural Development and Food Security International Working Paper Series Year, paper n. 14/1 Status of On –farm Water Use Efficiency and Source of Inefficiency in the Sudan Gezira Scheme, University of Pavia

Emmenegger, R. (2016). Decentralization and the Local Developmental State: Peasant Mobilization in Oromiya, Ethiopia. *Africa* 86(2): 263-287

ENA (2018) Ethiopia: Grand Ethiopian Renaissance Dam Bond Week to Begin On Sunday, 14 March 2018, http://www.ena.gov.et/en/index.php/economy/item/4426-grand-ethiopian-renaissance-dam-bond-week-to-begin-on-sunday accessed on 15 April 2018

ENPoE (2013) Ethiopian National Panel of Experts on the GERDP, Unwarranted Anxiety The Grand Ethiopian Renaissance Dam (GERD) and some Egyptian Experts Hyperbole- A Rebuttal of Statement made by Group of the Nile Basin (GNB) of Cairo University, June 2013; Ethiopioan Ministry of Water Resources, Irrigation and Energy Website: http://www.mowr.gov.et/index.php?pagenum=0.1&ContentID=88 last accessed 9 March 2016

ENTRO (2015) Eastern Nile Technical Regional Office Irrigation Toolkit

EPRDF (2007) 'Strategy of revolutionary democracy, tactics and the question of leadership', internal document, Addis Ababa: EPRDF (translated from Amharic).

Ertsen, M.W. (2006) Colonial irrigation. Myths of emptiness, Landscape Research, 31(2): 147–167. DOI : 10.1080/01426390600638588

Ertsen, M.W. (2016) Improvising planned development on the Gezira Plain, Sudan 1900-1980, Palgrave MacMillan, Houndmills

ESAT (2016) The Ethiopian Satellite Television & Radio (ESAT) News item 7 December 2016: Ethiopia: Sixty hectares of sugar plantation goes up in flames, https://ethsat.com/2016/12/ethiopia-sixty-hectares-sugar-plantation-goes-flames/ accessed 25-11-2017

ESC (Ethiopian Sugar Corporation) (2019) Comparative Description of Developments in the Sugar Industry Sector, http://www.ethiopiansugar.com/about/ Accessed on 5 May 2019

Escobar, A. (1995). Encountering Development: The Making and Unmaking of the Third World. Princeton: Princeton University Press.

ESIS (2014) Egyptian State Information Service (30 June 2014) Sisi, gov't discuss 4-million-feddan reclamation http://www.sis.gov.eg/Story/78577/Sisi%2c-gov%26rsquo%3bt-discuss-4-million-feddan-reclamation?lang=en-us, Accessed 10 March 2016

ESIS (2015a) Egyptian State Information Service (21 March 2015) 26 Egyptian, Arab companies start reclamation of 670,000 feddans http://www.sis.gov.eg/En/Templates/Articles/tmpArticleNews.aspx?ArtID=91522#.VuFezvkrLIU accessed 10 March 2016

ESIS (2015b) Egyptian State Information Service (29 September 2015) PM holds meeting on 4 mn feddans reclamation http://www.sis.gov.eg/En/Templates/Articles/tmpArticleNews.aspx?ArtID=96601#.VuFm4vkrLIU accessed 10 March 2016

ESIS (2015c) Egyptian State Information Service (31 December 2015) Irrigation Minister: Two thirds of 1.5-million-feddan reclamation project depends on subterranean water, http://www.sis.gov.eg/En/Templates/Articles/tmpArticleNews.aspx?ArtID=98676#.VuFkTfkrLIV accessed 10 March 2016

Esteva, G, Escobar, A (2017) Post-development @ 25: On 'being stuck' and moving forward, sideways, backward and otherwise. Third World Quarterly 38(12): 2559–2572.

Ethiopian Herald (2018) 20 March 2018 edition: Ethiopia: Office Collects Over 10 Billion Birr

Euroconsult, Sir Alexander Gibb and Partners, TCS (1982) Gezira Rehabilitation Project - Main Report

Evans (1994) History of Nile flows In: P.P. Howell and J.A. Allan, eds. The Nile: Sharing a Scarce Resource - A Historical and Technical Review of Water Management and of Economical and Legal Issues. Cambridge: Cambridge University Press, 1994, pp 27-63

Fahim, H. (1981) Dams, People and Development: The Aswan High Dam Case. New York: Pergamon press.

Faki (1982) Disparities in the management of resources between farm and national levels in irrigation projects, example of the Sudan Gezira Scheme, Agricultural Administration 9 pp 47-59

Fantini, E., Muluneh, T., Smit, H. (2018) Big projects, strong states? Large scale investments in irrigation and state formation in the Beles valley, Ethiopia, Chapter 5 in Menga, F., Swyngedouw, E., Eds. Water, Technology, and the Nation-State. Routledge Earthscan.

Fantini, E., Puddu, L. (2016) Ethiopia and international aid: development between high modernism and exceptional measures, chapter 4 in: Aid and Authoritarianism in Africa: Development without Democracy, 91-118

FAO (1986). *Ethiopian highland reclamation study*. Final Report, Vol 1. Rome: MoMFA.

FAO (1993) Water policies and agriculture, The State of Food and Agriculture, Rome

FAO (1997) Irrigation potential in Africa – a basin approach, Rome

FAO (1999) Transfer of irrigation management services - guidelines, by D. Vermillion & J.A. Sagardoy. Irrigation and Drainage Paper No. 58. Rome.

FAO (2000) New dimensions in water security, Water, society and ecosystem services in the 21st century FAO Land and Water Development Division Rome, 2000

FAO (2006) Fertilizer use by crop in the Sudan- Food and Agriculture Organization of the United Nations, Rome, Land and Plant Nutrition Management Service Land and Water Development Division

FAO (2007) Irrigation management transfer - Worldwide efforts and results, FAO Water Report 32, Rome

FAO (2011) Information Products for Nile Basin Water Resources Management – Synthesis Report FAO-Nile Basin Project GCP/INT/945/ITA 2004 to 2009, Rome

FAO (2015) Towards a Regional Collaborative Strategy on Sustainable Agricultural Water Management and Food Security in the Near East and North Africa Region, FAO RNE

FAO (2015b) Sudan Plan of action (2015-2019), FAO Regional Office for the Near East and North Africa

FAO (2016) Resilience analysis in Sudan - A policy Brief, Rome http://www.fao.org/3/a-i5594e.pdf

FAO (2018) Food and Agriculture Organisation Food import data Sudan over the 2014-2017 period retrieved from faostat.fao.org 20 December 2018

FAO (2018) WAPor The FAO portal to monitor Water Productivity, available at https://wapor.apps.fao.org/home/1, Accessed 17-05-2019

FAO and World Water Council (2018) 'Water accounting for water governance and sustainable development', White paper, Rome and Marseille, http://www.fao.org/3/i8868en/i8868en.pdf

FAOSTAT (2017) FAO Sudan statistical information on food availability obtained from http://www.fao.org/faostat/en/#country/276, accessed 12 December 2018

Farah Hasan Adam (1987) Evolution of the Gezira Patterns of Development within the Context of the History of Sudanese Agrarian Relations. in Elfatih Shaaeldin (ed.), The Evolution of Agrarian Relations in the Sudan: A Reader, The Hague.

Farbrother H.G., (1974) Irrigation practices in the Gezira - Cotton research report, Republic of the Sudan, 1970-71, Research Memoirs, No 89 Cotton Research Cooperation, London (pp36-99)

Feyerabend, P. (1981) Problems of Empiricism: Volume 2: Philosophical Papers

First, R. (1970) The Barrel Of A Gun: Political Power In Africa and the Coup D'Etat. London: The Penguin Press. 513 pp

Figueres, C., Schellnhuber, H. J., Whiteman, G., Rockström, J., Hobley, A., & Rahmstorf, S. (2017). Three years to safeguard our climate. Nature News, 546(7660), 593–595.

Floyd, R. and R.A. Matthew eds. (2013) Environmental Security: Approaches and Issues, Routledge, Abingdon.

Foucault, M. (1995[1977]). Discipline and punish: The birth of the prison (2nd edition), Vintage Books, Random House, New York

Foucault, M. (2007 [1978]) Security, Territory, Population: Lectures at the Collège de France 1977—1978, Translated by Graham Burchell, Picador Books, New York

Franco J., L. Mehta and G.J. Veldwisch (2013) 'The Global Politics of Water Grabbing', *Third World Quarterly* 34(9): 1651-75.

Frankl, A., Nyssen, N., De Dapper, M., Haile, M., Billi, P., Munro, R.N., Deckers, J., Poesen, J. (2011). Linking long-term gully and river channel dynamics to environmental change using repeat photography (Northern Ethiopia). *Geomorphology* 129: 238–51.

FT (Financial Times) (2014) Sudan tycoon's battle for growth – 12 February 2014 article retrieved on 5 February 2019 from https://www.ft.com/content/94a54080-9274-11e3-8018-00144feab7de

Gadkarim, Hassan Ali (2010) "Oil, Peace and Development: The Sudanese Impasse". ISSN 1890 5056. ISBN 978-82-8062-283-9. Peace building in Sudan: Micro-Macro Issues Research Programme, CMI.

Gaitskell, A. (1959). Gezira, A Story of Development in the Sudan (London: Faber, 1959)

Gandy, M. (2002). Concrete and clay: Reworking nature in New York City. Cambridge, MA: The MIT Press.

Garstin (1899) Note on the Soudan, Cairo

Garstin and Dupuis (1904) Report upon the basin of the Upper Nile, with proposals for the improvement of that river, Cairo.

Gebremichael, D., Nyssen, J., Poesen, J., Deckers, J., Mitiku Haile, Govers, G., Moeyersons, J. (2005). Effectiveness of stone bunds in controlling soil erosion on cropland in the Tigray highlands, Northern Ethiopia. *Soil Use & Management* 21: 287–97.

Gebremichael, T.G., Mohamed, Y.A., Betrie, G.D., Van der Zaag, P. & Teferi, E. (2013). Trend analysis of runoff and sediment fluxes in the Upper Blue Nile basin: A combined analysis of statistical tests, physically-based models and landuse maps. Journal of Hydrology 482: 57-68, DOI http://dx.doi.org/10.1016/j.jhydrol.2012.12.023

Georgakakos, A. and Yao, H. (2000) An Assessment of Development Options, Management Strategies, and Climate Scenarios for the Nile Basin, Georgia Water Resources Institute, Georgia Institute of Technology

Geressu, R.T. and Harou, J.J. (2015) Screening reservoir systems by considering the efficient trade-offs— informing infrastructure investment decisions on the Blue Nile, Environmental Resources Letters 10, pp1-13

Gertel, J., Rottenburg, R., & Calkins, S. (Eds.). (2014). Disrupting Territories: Land, Commodification and Conflict in Sudan. Boydell and Brewer.

Getaneh, Abiti (2011) Ethiopian Water Resources Potential & Development of Ethiopia, Ministry of Water & Energy, presentation accessed at 6 March 2016 http://www.partnersvoorwater.nl/wp-content/uploads/2011/11/PresentatieontwikkelingintergraalwaterbeheerinEthiopie.pdf

Gezira Scheme (2011) Cropped Area and Yield record for the Gezira scheme 1964-2010

Gezira scheme (2012) Gezira scheme excavation record 2010-2012

Ghazouani, W.; Molle, F.; Swelam, A.; Rap, E.; Abdo, A. 2014. Understanding farmers' adaptation to water scarcity: a case study from the western Nile Delta, Egypt. Colombo, Sri Lanka: International Water Management Institute (IWMI). 31p. (IWMI Research Report 160). doi: 10.5337/2015.200

Gismalla, Y.A. (2010) Review of the Sediment Monitoring Program in Gezira Scheme-Sudan, Journal of Science and Technology 11(2) 1-8

Gleick, P., Iceland, C. (2018) Water, Security, and Conflict. Issue Brief: World Resource Institute and Pacific Institute, Washington, D.C.

Goethe, J.W. (1987 [1776-1832]) Schriften zur Morphologie, ed. Dorothea Kuhn, vol. 24, Sämtliche Werke: Briefe, Tagebücher und Gespräche, Frankfurt am Main: Deutscher Klassiker Verlag, 1342pp

Goldman M (2005) Imperial Nature: The World Bank and the Making of Green Neoliberalism. New Haven, CT: Yale University Press

Goor Q, Halleux C, Mohamed Y and Tilmant A (2010) Optimal operation of a multipurpose multireservoir system in the Eastern Nile River Basin Hydrol. Earth Syste. Sci. 14 1895–908

GoE (Government of Ethiopia) (2010) Ethiopia's Growth and. Transformation Plan (GTP) for 2010/11−2014/15, Addis Ababa

GoS (Government of Sudan) (2008) Government of Sudan Agricultural policy, Khartoum

GoS (Government of Sudan) (2005) The Gezira Scheme Act 2005 – translated from Arabic, Government of the Republic of the Sudan, Khartoum

GoS (2015) Government of the Republic of Sudan: Ministry of Agriculture and Irrigation Ministry of Livestock, Fisheries and Rangelands Ministry of Environment, Forestry and Physical Development- Country Programming Framework for Sudan PLAN OF ACTION (2015-2019): Resilient Livelihoods for Sustainable Agriculture, Food Security and Nutrition

Goswami, M. (2011) Producing India: From Colonial Economy to National Economy, Chicago University Press, Chicago, 385 pages

Grey, D. & Sadoff, C. W. (2007) Sink or Swim? Water Security for Growth and Development. Water Policy, 9, 545-571.

Gross and Levitt (1994) Higher Superstition: The Academic Left and its Quarrels with Science

Gupta, J., Ahlers, R., & Ahmed, L. (2010). The Human Right to Water: Moving Toward Consensus in a Fragmented World. *Review of European Community & International Environmental Law, 19*(3), 294-305. https://doi.org/10.1111/j.1467-9388.2010.00688.x

GWP (Global Water Partnership) (2000) Towards water security: a framework for action. Stockholm, Sweden: GlobalWater Partnership.

GWP (Global Water Partnership) (2014a) Water Security: Putting the Concept into Practice, Global Water Partnership Technical Committee, TEC Background papers No. 20, Stockholm

GWP (Global Water Partnership) (2014b) Ecosystem Services and Water Security. Briefing Note. GWP, Stockholm, Sweden.

GWP (Global Water Partnership) (2015) Increasing Water Security: The Key to Implementing the Sustainable Development Goals. TEC Background Paper No 22. GWP, Stockholm, Sweden

Habteyes, B.G. , Hasseen El-bardisy, H.A.E, Amer, S.A., Schneider, V.R, Ward, F.A (2015) Mutually beneficial and sustainable management of Ethiopian and Egyptian dams in the Nile Basin, Journal of Hydrology 529, 1235–1246

Hajer, M. & Versteeg, W. (2018): Imagining the post-fossil city: why is it so difficult to think of new possible worlds?, Territory, Politics, Governance, DOI: 10.1080/21622671.2018.1510339

Hall 1993, Tatcherism today, New Statesman and Socieity, 26 November 1993

Hamdan, G (1972) Evolution of Irrigation Agriculture in Egypt 118-142 In: Stamp ed. A history of land use in Arid Regions, UNESCO- Paris

Hamed, E.M., Soberi, S, Elawad, O.M.A. (1992) Sudan Water cousmption from the Nile Waters - Ministry of Irrigation Sudan

Haraway, D. J. (1991) Simians, Cyborgs and Women. The Reinvention of Nature. London, U.K.: Free Association Books.

Haraway D.J. (1997) Modest Witness at Second Millennium. FemaleMan Meets OncoMouse: Feminism and Technoscience, Routledge, New York, London

Haraway, D.J. (2016) Staying with the Trouble: Making Kin in the Chthulucene. Durham, NC: Duke University Press

Haraway, D.J. and Harvey, D. (1995) Nature, politics, and possibilities: a debate and discussion with David Harvey and Donna Haraway. Environment and Planning D: Society and Space, 13: 507–527.

Harding, S.G. (1986) The Science Question in Feminism, Cornell University Press

Harding, S.G. (2015) Objectivity and Diversity, Another Logic of Scientific Research, University of Chicago Press, Chicago

Hardin, G. (1968). The Tragedy of the Commons. Science 162: 1243-1248

Harvey, D. (1982) The Limits to Capital. Oxford: Blackwell

Harvey, D. (2001) Spaces of Hope, University of California press, Berkeley and Los Angeles

Hereher, M.E. (2001) Mapping coastal erosion at the Nile Delta western promontory using Landsat imagery, Environ Earth Sci, 64: 1117. https://doi.org/10.1007/s12665-011-0928-9

Hewett, R.M.G. (1989) Pumped Irrigation on the White and blue Niles, Sudan. In: Rydzwewski and Wards eds. Irrigation Theory and practice, Proceedings of the interantational conference held at the University of Southampton 12-15 September, 1989, Pentech Press, London

Hoben, A. (1973). *Land tenure among the Amhara of Ethiopia: The dynamics of cognatic descent.* Chicago: University of Chicago Press.

Hoben, A. (1995). Paradigms and politics: The cultural construction of environmental policy in Ethiopia. *World Development* 23: 1007–21.

Homer-Dixon, T. F. (1994) Environmental Scarcities and Violent Conflict, International Security 19(1): 5-40.

Howell, P.P. and Allan, J.A., editors, (1994) The Nile: sharing a scarce resource. Cambridge: Cambridge University

HRC (Hydraulics Research Centre) (2011) Data record sediment concentration measurement in the Gezira main canal 1995-2011, Wad Medani, Sudan

HRC (Hydraulics Research Center Sudan) (2016) Water Productivity in the Eastern Nile: Case Studies from Sudan, Egypt and Ethiopia, December 2016

HRC (Hydraulics Research Center Sudan) (2018) Satellite based ICT for improved crop production in the Gezira irrigation scheme in Sudan, http://wre.gov.sd/hrc/index.php/portfolio/project-3/, accessed 20-12-2018

Hurni, H., Tato, K., Zeleke, G. (2005). The implications of changes in population, land use, and land management for surface runoff in the Upper Nile Basin area of Ethiopia. *Mountain Research and Development* 25: 147–54.

Hurst H.E., Black R.P., Simaika Y.M. (1946) The Nile basin, vol. VII. The future conservation of the Nile. Cairo, Egypt: Government Press.

Hurst, H.E. (1957) *The Nile: A General Account of the River and the Utilization of its Waters.* London: Constable.

IBRD (1959) Report of the technical mission on Sudan Irrigation, Department of Technical Operations, IBRD

IBRD (1963) The Ten Year Plan of Economic and Social Development of the Sudan (1961/62-1970/71), Department of Operations, IBRD, Africa

IBRD (1968) Mechanised farming project, Sudan, Report nr TO 434a

IBRD (1972) Appraisal of second Mechanized farming project, Sudan, Report nr PA 126a

IBRD (1966) Sudan mission report, Leonard. B. Rist Chief of Mission (Ed.) Main report.

IBRD (International Bank for Development and Reconstruction) (1968) Mechanized Farming Project, Sudan

IBRD (International Bank for Development and Reconstruction) (1982) The Gezira scheme Study Mission Final Report Vol IV, Annex E.

ICARDA (2011) Water and Agriculture in Egypt, Technical paper based on the Egypt-Australia-ICARDA Workshop on On-farm Water-use Efficiency, International Center for Agricultural Research in the Dry Areas, Cairo

ICG (International Crisis Group) (2013) Sudan's Spreading Conflict (II): War in Blue Nile, Africa Report N°204

IMF (2014) 2014 ARTICLE IV consultation and second review under staff-monitored program Sudan - IMF Country Report No. 14/364

IMF (2017) Staff report for the 2017 Article IV consultation – Debt sustainability analysis – Sudan https://www.imf.org/external/pubs/ft/dsa/pdf/2017/dsacr17364.pdf

Ingold, T. (2017) Anthropology contra ethnography, Hau: Journal of Ethnographic Theory 7 (1): 21–26

Ismail, A. (2009) Private Equity and Venture Capital in Emerging Markets: A Case Study of Egypt and the MENA Region, PhD dissertation MIT.

IWMI (2014) On target for people and planet: setting and achieving water related sustainable development goals. Van der Bliek, J., McCornick, P. & Clarke, J. (eds). Sri Lanka, International Water Management Institute.

James, W., Donald L. Donham, Eisei Kurimoto, Alessandro Triulzi eds. (2002) Remapping Ethiopia: Socialism and After. Athens: Ohio University Press

Kahsay, T. (2017) 'Towards Sustainable Water Resources Management in the Nile River Basin: A Global Commutable General Equilibrium Analysis', PhD thesis Vrije Universiteit Amsterdam

Kahsay, T.N., Kuik, O., Brouwer, R. and Van der Zaag, P. (2015) Estimation of the transboundary economic impacts of the Grand Ethiopia Renaissance Dam: A computable general equilibrium analysis Water Resources and Economics 10 (2015 )14–30

Kaika, M. (2005) City of Flows. London and New York: Routledge.

Kaika, M. (2017). 'Don't call me resilient again!': the New Urban Agenda as immunology … or … what happens when communities refuse to be vaccinated with 'smart cities' and indicators. Environment and Urbanization, 29(1), 89–102.

Kalinovsky, A. M. (2013). Not Some British Colony in Africa: The Politics of Decolonization and Modernization in Soviet Central Asia, 1955-1964. Ab Imperio 2013(2), 191-222.

Kamski, B. (2016) The Kuraz Sugar Development Project (KSDP) in Ethiopia: between 'sweet visions' and mounting challenges, Journal of Eastern African Studies, 10:3, 568-580, DOI: 10.1080/17531055.2016.1267602

Karimi, P. and W.G.M. Bastiaanssen, (2015) Spatial Evapotranspiration, Rainfall and Land Use Data in Water Accounting - Part 1: Review of the accuracy of the remote sensing data, HESS 19, 507–532

Karimi, P., Molden, D., Notenbaert, A., and Peden, D.(2012) Nile basin farming systems and productivity. In The Nile river basin; water, agriculture, governance and livelihoods. Edited by Awulachew, et al. pp. 133 -153, Routledge, UK.

Kautsky, K. (1988[1899]). The Agrarian Question, 2 volumes. London: Zwan Publications. First published in 1899.

Keeley, J., Scoones, I. (2000). Knowledge, power and politics: The environmental policy-making process in Ethiopia. *The Journal of Modern African Studies* 38: 89–120.

Keeley, J., W. Michago Seide, A. Eid and A. Lokaley Kidewa (2014) Large-scale Land Deals in Ethiopia: Scale, Trends, Features and Outcomes to Date. London: IIED.

Kemerink, J.S., L. E. Mendez, R. Ahlers, P. van der Zaag (2013) The question of inclusion and representation in rural South Africa: challenging the concept of Water User Associations as a vehicle for transformation. Water Policy 15: 243–257

Kemerink, J.S., S.N. Munyao, K. Schwartz, R. Ahlers, P. van der Zaag (2016) Why infrastructure still matters: unravelling water reform processes in an uneven waterscape in rural Kenya. International Journal of the Commons 10(2): 1055–1081.

Keulertz, M. (2016) Inward investment in Sudan: the case of Qatar. In E. Sandström, A. Jägerskog, & T. Oestigaard (Eds.), Land and hydropolitics in the Nile River Basin: Challenges and new investments. Abingdon: Routledge-Earthscan. pp73-88

Kirsch, S. (2015). Cultural geography III: Objects of culture and humanity, or, re- 'thinging' the Anthropocene landscape. *Progress in Human Geography*, 39(6), 818-826

Kooy, M. and Bakker, K. (2008) Technologies of Government: Constituting subjectivities, spaces, and infrastructures in colonial and contemporary Jakarta. *International Journal of Urban and Regional Research*, 32(2):375-391.

Kubota et al (2017) Water and salt movement in Soils of the Nile Delta, Chapter 7 in Irrigated Agriculture in Egypt: Past, Present and Future, Eds Masayoshi Satoh, Samir Aboulroos, Springer, Cham Switzerland.

Lakew Desta, Carruci, V., Asrat Wendem-Agenehu, Yitayew Abebe (eds.) (2005). Community based participatory watershed development: A guideline. Addis Ababa: Ministry of Agriculture and Rural Development.

Lanckriet, S., Derudder, B., Naudts, J., Bauer, H., Deckers, J., Haile, M., Nyssen, J. (2015). A political ecology perspective of land degradation in the North Ethiopian Highlands. *Land Degradation and Development* 26: 521–30.

Lankford, B.; Bakker, K.; Zeitoun, M.; Conway, D. (Eds.). (2013) Water security: principles, perspectives and practices. London, UK: Routledge. pp.148-165.

Large, D and Patey, L.A. (2011) Sudan Looks East, China, India & the politicls of Asian alternatives, African issues, James Currey, Rochester

Latour, B. (1987) Science in Action. Cambridge MA: Harvard University Press

Latour, B. (1993) We have never been modern, Cambridge MA: Harvard University Press

Latour, B. (2005). Reassembling the social: an introduction to actor-network-theory, Oxford: Oxford University Press.

Latour, B. (2018) Down to Earth: Politics in the New Climatic Regime, Polity

Law, J. (2004) After Method: Mess in Social Science Research, Routledge, London

Lefort, R. (2012) Free market economy, 'developmental state' and party–state hegemony in Ethiopia: The case of the 'model farmers'. *The Journal of Modern African Studies* 50: 681–706. doi:10.1017/S0022278X12000389

Li, T.M. (2007) The will to improve, Duke University Press

Li, T.M. (2007). *The will to improve: Governmentality, development, and the practice of politics.* Durham, NC: Duke University Press.

Link, P. M., Scheffran, J., Ide, T., (2016) Conflict and cooperation in the water-security nexus: a global comparative analysis of river basins under climate change, WIREs Water, 3:495–515. doi: 10.1002/wat2.1151

Linnaeus, C. (1737) Genera plantarum, translated introduction by Staffan Müller-Wille and Karen Reeds, Studies in History and Philosophy of Biological and Biomedical Science 38 (2007): 563–72

Linton, J. (2010) What is Water? The history of a Modern Abstraction, UBC press 2010

Litfin, K. (1997) The Gendered Eye in the Sky: A Feminist Perspective on Earth Observation Satellites, Frontiers: A Journal of Women Studies, Vol. 18, No. 2, pp. 26-47

Liu, X. (2014 Open-Channel Hydraulics: From Then to Now and Beyond, Chapter 2 in Handbook of Environmental Engineering, Volume 15: Modern Water Resources Engineering Edited by: L.K. Wang and C.T. Yang, DOI 10.1007/978-1-62703-595-8_2, Springer Science+Business Media New York 2014

Lorde, A. (1983) 'The Master's Tools Will Never Dismantle the Master's House." In This Bridge Called My Back: Writings by Radical Women of Color, ed. Cherrié Moraga and Gloria Anzaldúa. 98-101. New York: Kitchen Table

LUA 2019 Live Universal Awareness map https://sudan.liveuamap.com/en/2019/12-february-demonstration-stand-for-staff-of-dal-food-company accessed 18-02-2019

MacDonald, M. (1920) Nile Control, Government Press, Cairo

Mann, L. (2013) 'We do our bit in our own space': DAL Group and the development of a curiously Sudanese enclave economy. Journal of Modern African Studies 51, (2) pp. 279-303

Manning R. (1895). On the flow of water in open channels and pipes - Supplement to a paper read on the 4th December 1889, published in the Transactions, 1891, vol. XX, p. 161. Transactions of the Institution of Civil Engineers of Ireland, 24, 179-207.

Marx, K. (2007 [1867], Capital: A Critique of Political Economy, Vol 1. New York: Cosimo classics economics

Massey, D. (1993). Power geometry and a progressive sense of place. In J. Bird, B. Curtis, T. Putnam, G. Robertson, & L. Tickner (Eds.), Mapping the futures (pp. 59–69). London: Routledge.

Massey, D. (1994) Space, Place and Gender. Cambridge: Polity Press.

Massey, D. (2005) For Space. London: Sage

Matthews, N., Nicol, A., & Seide, W. (2013). Constructing a new water future? An analysis of Ethiopia's current hydropower development. In J. A. Allan, M. Keulertz, S. Sojamo, & J.Warner (Eds.), Handbook of land and water grabs in Africa (pp. 311–323). London: Routledge.

McMichael, P. (1990). Incorporating Comparison within a World-Historical Perspective: An Alternative Comparative Method. American Sociological Review, 55(June), 385-397.

McMichael, P. (1992) Rethinking Comparative Analysis in a Post-developmentalist Context. International Social Science Journal, 133, 350-365.

McMichael, P. (1997) 'Rethinking Globalization: the agrarian question revisited.Review of International Political Economy,4 (4): 630–62

McMichael, P. (2012) Development and social change – A Global Perspective 5th edition – Sage publications

Meehan, K.M. (2014). Tool-power: Water infrastructure as wellsprings of state power. *Geoforum* 57: 215–24. http://dx.doi.org/10.1016/j.geoforum.2013.08.005

Mehta L., G.J. Veldwisch and J. Franco (eds.) (2012) Special Issue: 'Water grabbing? Focus on the (re)appropriation of finite water resources', Water *Alternatives* 5(2): 193-542

Merme, V., Ahlers, R., Gupta, J. (2014) Private equity, public affair: Hydropower financing in the Mekong Basin, Global Environmental Change 24, 20-29

MIH (Ministry of Irrigation and Hydro-Power, Sudan) (1934) Gezira Canal Regulation Handbook

Mikhail, A (2011). *Nature and Empire in Ottoman Egypt: an environmental history.* Cambridge: Cambridge University Press

Mitchell, D. (2012). *They saved the crops – Labor, landscape, and the struggle over industrial farming in Bracero-era California.* Athens and London: The University of Georgia Press.

Mitchell, T. (1999) Society, Economy, and the State Effect . In State/Culture: State-Formation after the Cultural Turn . George Steinmetz , ed. Pp. 76 – 97 . Ithaca , NY : Cornell University Press .

Mitchell, T. (2002). *Rule of experts: Egypt, techno-politics, modernity.* Berkeley: University of California Press.

Mitchell, T. (2009): Carbon democracy, Economy and Society, 38:3, 399-432

MoA (Ministry of Agriculture Ethiopia) (2013). *Sustainable land management program (SLMP) integrated implementation progress report for the time period 8th July 2012–8th January 2013.* Addis Abba: Ministry of Agriculture.

MoA (Ministry of Agriculture Sudan) (2008) Sudan plan for agricultural reform.

Mohamed, A.S. (2001) Water Demand Management: Approach, Experience and Application to Egypt. PhD dissertation. Delft: IHE Delft and Delft University of Technology.

Mohamed, Y. A. (2005). *The Nile hydroclimatology: impact of the Sudd wetland.* London: Taylor & Francis.

Mohammed, A.E., Ahmed, A.E., Ahmed, A., 2008. Food Security in Sudan: Policies and Strategies. Globalization, Technology and Sustainable Development Book Series, Inderscience Enterprises Limited, UK. ISBN (Print): 0-907776–36–1.

MoI (Ministry of Irrigation Egypt) (1982) Egypt Water Use and Management Project, Cairo and Ford Collins

MoI (Ministry of Irrigation Egypt), UNDP, IBRD (1980) Water Master Plan - Water Demands - Arab Republic of Egypt

MoI Egypt, UNDP, IBRD (1980) Water Master Plan - Water Quality - Arab Republic of Egypt

MoI Egypt, UNDP, IBRD (1981) Water Master Plan - Main report - Arab Republic of Egypt

MoIWR (2010) Ministry of Irrigation and Water Resources Sudan, canal excavation records

MoIWR (2012) Ministry of Irrigation and Water Resources Sudan, records of diversion of water at Sennar

Mol, A. (2002). *The body multiple: Ontology in medical practice*. Durham, NC, and London: Duke University Press.

Molden, D. (1997) Accounting for water use and productivity, SWIM Paper 1, International Irrigation Management Institute, Colombo, Sri Lanka.

Molle, F. (2006) Planning and managing water resources at the river-basin level: Emergence and evolution of a concept. Colombo, Sri Lanka: International Water Management Institute. 38p. (IWMI Comprehensive Assessment Research Report 16)

Molle, F. (2009) River-basin planning and management: The social life of a concept, Geoforum 40 (2009) 484–494

Molle, F. and Wester, P. (2009) River basin trajectories: an inquiry into Changing Waterscapes, Ch 1 in Molle, F. And Wester, P. (eds) River basin trajectories : societies, environments and development, MPG Books Group, Bodmin

Molle, F., Mollinga, P., & Wester, F. (2009). Hydraulic bureaucracies and the hydraulic mission: Flows of water, flows of power. Water Alternatives, 3(2), 328–349.

Molle, F., Gaafar, I., El-Agha, D. E., Rap, E. (2018) The Nile Delta's water and salt balances and implications for management. Agricultural Water Management, 197:110-121. doi: 10.1016/j.agwat.2017.11.016

Mollinga, P.P. (2003). On the waterfront: water distribution, technology and agrarian change in a South Indiancanal irrigation system. Orient Longman Private Limited: New Delhi

Mollinga, P.P. (2014) Canal irrigation and the hydrosocial cycle: The morphogenesis of contested water control in the Tungabhadra Left Bank Canal, South India, Geoforum, (57)pp 192-204 doi: 10.1016/j.geoforum.2013.05.011.

Mollinga, P. P., and A. Bolding (1996) Signposts of struggle. In Crops, people and irrigation, ed. G. Diemer, and F. P. Huibers. London: Intermediate Technology Publications.

Mollinga, P. P., and A. Bolding (2004) The Politics of Irrigation Reform: Contested Policy Formulation and Implementation in Asia, Africa and Latin America, Ashgate

Mollinga, P.P. and Veldwisch, G.J. (2016). Ruling by canal: Governance and system-level design characteristics of large-scale irrigation infrastructure in India and Uzbekistan. Water Alternatives 9(2): 222-249

Monsieurs, E., Dessie, M., Poesen, J., Deckers, J., Verhoest, N., Nyssen, J., Adgo, E. (2015b). Seasonal surface drainage of sloping farmland and its hydrogeomorphic impacts. *Land Degradation and Development*. http://dx.doi.org/10.1002/ldr.2286.

Monsieurs, E., Poesen, J., Dessie, M., Adgo, E., Verhoest, N.E.C., Deckers, J., Nyssen, J. (2015a). Effects of drainage ditches and stone bunds on topographical thresholds for gully head development in North Ethiopia, *Geomorphology*, 234, 193-203

Moore, J.W. (2015) Capitalism in the web of life, Verso

Mosse, D. (1997) The symbolic making of a common property resource: History, ecology and locality in a tank-irrigated landscape in South India, Development and Change 28(3):467–504

NBE 2018 Ethiopia approves 12.8 billion national budget, 6 July 2018, https://newbusinessethiopia.com/ethiopia-approves-12-8-billion-national-budget/ accessed 31 July 2018

NBI (2009) Transboundary Benefit Sharing Framework, Nile Basin Initiative Secretariat, Entebbe

NBI (2012) State of the Nile River Basin Report, Nile Basin Initiative Secretariat, Entebbe

NBI (2014) Nile Basin Initiative, Corporate Report 2014, Entebbe

NBI (2016) Nile Basin Water Resources Atlas, Nile Basin Initiative (NBI), New Vision Printing, Kampala, Uganda, p201

NBI-ENTRO (2013) First Joint Multipurpose Program Identification (JMP1 ID) - Strategic Perspectives and options assessment on Blue Nile Multipurpose development, NBI-ENTRO, Addis Ababa

Newhouse (1939) The Training of the Upper Nile, Sir Isaac Pitman and Sons, London

Niblock, T. (1987) Class and Power in Sudan The Dynamics of Sudanese Politics, 1898-1985, Albany, State University of New York Press.

Nicol, A. & A. E. Cascão (2011) Against the flow–new power dynamics and upstream mobilisation in the Nile Basin. Review of African Political Economy, 38, 317-325.

Norgaard, R.B. (2010) Ecosystem services: from eye-opening metaphor to complexity blinder. Ecological Economics 69, 1219–1227.

Norman, E.S., Cook, C., Cohen, A., EDS (2015) Negotiating Water Governance: Why the Politics of Scale Matter, Routledge

Nyssen, J., Poesen, J., Desta Gebremichael, Vancampenhout, K., D'aes, M., Gebremedhin Yihdego, Govers, G., Leirs, H., Moeyersons, J., Naudts, J., Nigussie Haregeweyn, Mitiku Haile, Deckers, J. (2007). Interdisciplinary on-site evaluation of stone bunds to control soil erosion on cropland in Northern Ethiopia. *Soil and Tillage Research* 94: 151–63.

O'Brien, J. (1981) Sudan: An Arab Breadbasket?. *MERIP Reports*, 99, 20-26.

O'Brien, J., (1983) The formation of the agricultural labour force in Sudan, Review of African Political Economy, no.26, pp15-34

O'Brien, J. (1984) The social reproduction of tenant cultivators and class formation in the Gezira Scheme, Sudan, Research in Economic Anthropology, Volume 6, pp217-241

O'Brien, J (1984a) The political economy of semi-proletarianisation under colonialism: Sudan 1925-50, in B. Munslow and B. Finch (eds.), Proleterianism in the Third World, Sydney, Croom Helm, 121-47

Obertreis, J.; Moss, T.; Mollinga, P. and Bichsel, C. 2016. Water, infrastructure and political rule: Introduction to the Special Issue. Water Alternatives 9(2): 168-181

Ojulu, O.M. (2013) Large-scale land acquisitions and minorities/indigenous peoples' rights under ethnic federalism in Ethiopia. PhD Thesis, Department of Peace Studies, University of Bradford.

Ong A. (2006) Neoliberalism as Exception: Mutations in Sovereignty and Citizenship. Durham, NC: Duke University Press.

Osman, I.S.E. (2015) Impact of Improved Operation and Maintenance on Cohesive Sediment Transport in Gezira Scheme, Sudan, PhD thesis, UNESCO-IHE and Wageningen University

Osman, M., Sauerborn, P. (2001). Soil and water conservation in Ethiopia. *Journal of Soils and Sediments* 1(2) 117–23.

Osman, R, Ferrari, E, McDonald, S. (2016) Water scarcity and irrigation efficiency in Egypt, Water Economics and Policy 2(4) 165-174

Ostrom, E. (1990) Governing the commons: The evolution of institutions for collective action. New York: Cambridge University Press.

Ostrom, E. (1992). Crafting institutions for self-governing irrigation systems. *Institute for Contemporary Studies*, 1-39.

Owen, R. (1969) *Cotton and the Egyptian Economy, 1820 – 1914: A Study in Trade and Development* (Oxford: Clarendon,)

Owen, R. (2002) "From Liberalism to Liberal Imperialism: Lord Cromer and the First Wave of Globalization in Egypt." In *Histories of the Modern Middle East: New Directions*, edited by Israel Gershoni, Hakan Erdem, Ursula Wokock, 95-112. London: Lynne Rienner Publishers, 2002.

Oya, C (2013) Methodological reflections on land "grab" databases and the land "grab" literature"rush", Journal of Peasant Studies, 40(3), pp 503–520.

Pahl-Wostl C. (2006) Transitions towards adaptive management of water facing climate and global change. Water Resour. Manage. 21, 49–62. (doi:10.1007/s11269-006-9040-4)

Pahl-Wostl, C., J. Sendzimir, P. Jeffrey, J. Aerts, G. Berkamp, and K. Cross (2007a) Managing change toward adaptive water management through social learning. Ecology and Society 12(2): 30. [online] URL: http://www.ecologyandsociety.org/vol12/iss2/art30/.

Pahl-Wostl, C., M. Craps, A. Dewulf, E. Mostert, D. Tabara, and T. Taillieu. (2007b) Social learning and water resources management. Ecology and Society 12(2): 5. [online] URL: http://www.ecologyandsociety.org/vol12/iss2/art5/

Pahl-Wostl, C., Lebel, L., Knieper, C., Nikitina, E. (2012) From applying panaceas to mastering complexity: Toward adaptive water governance in river basins, In Environmental Science & Policy, Volume 23, 2012, Pages 24-34, ISSN 1462-9011, https://doi.org/10.1016/j.envsci.2012.07.014.

Pahl-Wostl, C., Lebel, L., Knieper, C., Nikitina, E. (2012) From applying panaceas to mastering complexity: Toward adaptive water governance in river basins, In Environmental Science & Policy, Volume 23, 2012, Pages 24-34, ISSN 1462-9011, https://doi.org/10.1016/j.envsci.2012.07.014.

Pahl-Wostl, C., Bhaduri, A. & Gupta, J. (eds.) (2016) Handbook on water security, Edward Elgar

Pankhurst, A. and Piguet, F (2009) Moving People in Ethiopia: Development, Displacement and the State. Oxford: James Currey,

Patey. L.A. (2007) State rules: oil companies and armed conflict in Sudan, Third World Quarterly, 28(5) pp 1009-10

Peel, S. (1904) The binding of the Nile and the New Soudan, Negro Universities Press, New York

People's Republic of China (2013) China-Africa Economic and Trade Cooperation, Information Office of the State Council, The People's Republic of China, Beijing

PJTC (1959) Permanent Joint Technical Commission for Nile Waters, Agreement Between the Republic of the Sudan and the United Arab Republic of the Full Utilisation of the Nile Waters, Al Shaab printing house

Planel, S. (2014). A view of a bureaucratic developmental state: local governance and agricultural extension in rural Ethiopia, *Journal of Eastern African Studies* 8 (3): 420–37

Radwan, S. (1978) Agrarian Reform and Rural poverty: Egypt , 1952-1975. ILO, Geneva

Raffles, H. (2003) In Amazonia, Princeton University Press

Rap, E. & Wester, P. (2017): Governing the water user: experiences from Mexico, Journal of Environmental Policy & Planning, DOI: 10.1080/1523908X.2017.1326305

Reuters (2018) Sudanese hit by bread shortages as currency crunch escalates, online article 14 August 2018 available at https://www.reuters.com/article/sudan-economy/sudanese-hit-by-bread-shortages-as-currency-crunch-escalates-idUSL5N1V56LJ

Rhodes, R.A.W. (1996) The new governance: governing withoutgovernment. Polit Stud., XLIV: 652–667.

Richards, A. (1982) *Egypt's Agricultural Development, 1800-1980: Technical and social change.* Boulder, Colorado: Westview

Ritzema, H. (2009) Drain for gain: making water management worth its salt: subsurface drainage practices in irrigated agriculture in semi-arid and arid regions, UNESCO-IHE and Wageningen University

Robbins, P. (2007). *Lawn people: How grasses, weeds, and chemicals make us who we are.* Philadelphia: Temple University Press.

Robbins, P. (2012). *Political ecology: A critical introduction.* 2nd ed. Chichester, UK: John Wiley.

Rogers, P., R. Bhatia, and A. Huber (1997) Water as a Social and Economic Good: How to Put the Principle into Practice. TAC Background Papers No.2. Global Water Partnership,

Roy, A. (1999). The Greater Common Good. India Book Distributor, Bombay

Saad, R.(2002) Egyptian politics and the tenancy law. In R Bush (ed) Counter-Revolution inEgypt's Countryside (pp 103–125). London: Zed Books

Said, E.W. (1978) Orientalism, Routledge and Kegan Paul Ltd.

Salam, A. (2009) 'The Abdus Salam Report' on operation and maintenance in the Gezira

Salih Ahmed Ali, Yasir (2014): The Impact of Soil Erosion in the Upper Blue Nile on Downstream Reservoir Sedimentation , Delft  28 October 2014, PhD thesis UNESCO-IHE and TU-Delft

Salman M. A. Salman (2016) The Grand Ethiopian Renaissance Dam: the road to the declaration of principles and the Khartoum document, Water International, 41:4, 512-527, DOI: 10.1080/02508060.2016.1170374

Sandström, E. (2016) Dealing with water: emerging land investments and the hydropolitical landscape of the Nile Basin. In E. Sandström, A. Jägerskog, & T. Oestigaard (Eds.), Land and hydropolitics in the Nile River Basin: Challenges and new investments. Abingdon: Routledge-Earthscan. Pp14-35

Sauer, C. O. (1925). The morphology of landscape. *University of California Publications in Geography* 2(2): 19–53.

Savelli, E., Schwartz, K. and Ahlers, R. (2018) "The Dutch Aid and Trade Policy: Policy Discourses versus Development Practices in the Kenyan WASH Sector", Environment and Planning C: Politics and Space. https://doi.org/10.1177/0263774X18803364

Savenije, H.H.G., and Van Der Zaag, P. (2002). Water as an Economic good and demand management; paradigms with pitfalls. Water International 27(1): 98-104.

Schewe (2012) Website: https://ericschewe.wordpress.com/2012/07/07/district-map-of-the-presidential-election-in-lower-egypt-an-environmental-history/ last accessed 8 March 2016

SCRP (Soil Conservation Research Program) (2000). *Area of Anjeni, Gojam, Ethiopia: Long-term monitoring of the agricultural environment 1988–1994.* Bern: University of Bern, Centre for Development and Environment.

Segers, K., Dessein, J., Hagberg, S., Develtere, P., Haile, M. Deckers, J. (2009). Be like bees – The politics of mobilizing farmers for development in Tigray, Ethiopia. *African Affairs* 108, 91–109.

Shaaeldin, E. and Brown, R.(1988) In: Barnett, T and Abdelkarim, A. (eds.) (1988) Sudan: State, Capital and Transformation, Croom Helm, Beckenham, Kent. pp121 - 141

Shiferaw, B., Holden, S. (1999). Soil erosion and smallholders' conservation decisions in the highlands of Ethiopia. *World Development* 27(4): 739–52.

SID (Sudan Irrigation Department) (1943) Drawing SID 42/1 - 879, Silt content of canal water D.S. main regulator 1925-1938 samples, Wad Medani, Sudan

Siddig, K.H.A. and Babiker, B.I. (2011) Agricultural Efficiency Gains and Trade Liberalization in Sudan, Agricultural Economics Working Paper Series Working Paper No. 1

Siddig, K.H.A. and Grethe, H. (2015) Wheat Import Subsidies in the Sudan: Problems and Alternative Policy Options for Poverty Alleviation, Conference paper presented at the 18th Annual Conference on Global Economic Analysis, Melbourne, Australia

Sims, D. (2015) Egypt's Desert Dreams : Development or Disaster? Cairo: The American University in Cairo Press

SMC (2017) Sudanese Media Center 26 October 2017 - Chinese Company to Increase Investment in Sudan Cotton Industry, http://smc.sd/en/chinese-company-increase-investment-sudan-cotton-industry/ Accessed on 5 July 2018

SMEC (2012) RDHP Reservoir Operation Study – Final draft 2012

Smith, D. (1990). The conceptual practices of power: A feminist sociology of knowledge. Toronto: University of Toronto Press

Smith, N. (1984), Uneven Development -Nature, Capital, and the Production of Space, Third Edition (2011), The University of Georgia Press, Athens and London

Sneddon, C., Fox, C., 2011. The Cold War, the US Bureau of Reclamation, and the technopolitics of river basin development, 1950–1970. Political Geography 30 (8), 450–460.

Sneddon, C. (2012) The 'sinew of development': Cold War geopolitics, technical expertise, and water resource development in Southeast Asia, 1954–1975. Social Studies of Science 42(4): 564–590.

Sneddon, C. (2015) Concrete revolution - Large dams, cold war geopolitics and the US Bureau of Reclamation, The University of Chicago press, Chicago and London

Soliman, N.F. & Yacout, D.M.M. (2016) Aquaculture International (2016) 24: 1201. https://doi.org/10.1007/s10499-016-9989-9

Sowers, J. (2011). Remapping the Nation, Critiquing the State: Narrating Land Reclamation for Egypt's "New Valley'. Environmental Imaginarles of the Middle East. D. Davis and T. Burke. Athens, Ohio: Ohio University Press

Springborg, R. (1990) Rolling Back Egypt's Agrarian Reform, Middle East Report 166, pp. 28–38.

Srinivasan, V., Konar, M., & Sivapalan, M. (2017). A dynamic framework for water security. Water Security, 1, 12-20. https://doi.org/10.1016/j.wasec.2017.03.001

Ståhl, M. (1990). Environmental degradation and political constraints in Ethiopia. *Disasters* 14: 140–50.

Star, S. L. (1990), Power, technology and the phenomenology of conventions: on being allergic to onions. The Sociological Review, 38: 26–56. doi:10.1111/j.1467-954X.1990.tb03347.x

Strathern, M. ed. (2000) Audit cultures, Routledge, New York

Subramanian, A, Bridget Brown, and Aaron Wolf (2012) Reaching Across the Waters: Facing the Risks of Cooperation in International Waters. Washington DC: The World Bank Press.

Sudan Cotton Company (2014) Company profile http://www.sudan-cotton.com/profile.html accessed 4 March 2014

Sudan Gezira Board (1951) A Hand Book for New Personnel, McCorquodale, Khartoum

Sudan Tribune (2010)   Gezira investment by Egyptians off

Sudan tribune (2013)   12 March 2013 Sudan's Gezira agricultural project is failing, governor says. Accessed 5 December 2013 http://www.sudantribune.com/spip.php?article45813

Sudan Tribune (2015) Saudi Arabia to provide 1.7 billion dollars for Sudan's dam projects Accessed 6 March 2016 http://www.sudantribune.com/spip.php?article56960

Suhardiman, D. (2013) The power to resist:Irrigation management transfer in Indonesia.Water Alternatives 6(1): 25-41

Sultana F, Loftus A (eds). (2012) The right to water: politics, governance and social struggles. London, UK: Earthscan.

Sutcliffe, J. V. and Parks, Y. (1999) *The hydrology of the Nile*. IAHS Special Publication 5. Wallingford: IAHS Press.

Swyngedouw, E. (1999) 'Modernity and hybridity: nature, regeneracionismo, and the production of the Spanish waterscape, 1890–1930', Annals of the Association of American Geographers 89, pp. 443–465.

Swyngedouw, E. (2007) Technonatural revolutions: the scalar politics of Franco's hydro-social dream for Spain, 1939–1975. Transactions of the Institute of British Geographers NS 32 (1), 9–28.

Swyngedouw, E. (2014) 'Not a Drop of Water .....': State, Modernity and the Production of Nature in Spain, 1998-2010. 20(1): 67-92

Swyngedouw, E. (2015) Liquid Power: Contested Hydro-Modernities in Twentieth-Century Spain, MIT Press, Cambridge,

Taye, G., Poesen, J., Van Wesemael, B., Vanmaercke, M., Teka, D., Deckers, J., Goosse, T., Maetens, W., Nyssen, J., Hallet, V., Haregeweyn, N. (2013). Effects of land use, slope gradient, and soil and water conservation structures on runoff and soil loss in semi-arid Northern Ethiopia. *Physical Geography* 34 (3), 236–259.

Teferi Abate Adem (2012). The Local Politics of Ethiopia's Green Revolution in South Wollo. *African Studies Review* 55 (3) 81–102. doi:10.1017/S0002020600007216.

Teferi, E. (2015). Soil hydrological impacts and climatic controls of land use and land cover changes in the Upper Blue Nile (Abay) basin, PhD thesis, UNESCO-IHE Institute for Water Education

Teferi, E., Bewket, W., Uhlenbrook, S., Wenninger, J. (2013). Understanding recent land use and land cover dynamics in the source region of the Upper Blue Nile, Ethiopia: Spatially explicit statistical modeling of systematic transitions. *Agriculture, Ecosystems & Environment* 165: 98–117. doi: http://dx.doi.org/10.1016/j.agee.2012.11.007

Tekelab, S.G. (2015): Understanding catchment processes and hydrological modelling in the Abay/Upper Blue Nile basin, Ethiopia , PhD Thesis UNESCO-IHE and TU-Delft

Tekleab, S., Mohamed, Y.A., Uhlenbrook, S., Wenninger, J. (2014). Hydrologic responses to land cover change: The case of Jedeb Meso-scale catchment, Abay/Upper Blue Nile basin, Ethiopia. *Hydrological Processes* 28(20), 5149–61. doi: 10.1002/hyp.9998

Temesgen, M., S. Uhlenbrook, B. Simane, P. van der Zaag, Y. Mohamed, J. Wenninger and H.H.G. Savenije (2012). Impacts of conservation tillage on the hydrological and agronomic performance of Fanya juus in the upper Blue Nile (Abbay) river basin. Hydrol. Earth Syst. Sci. 16, 4725-4735; DOI: http://dx.doi.org/10.5194/hess-16-4725-2012

Tesfaye, A. (2013) Institutional-Economic Incentives for Sustainable Watershed Management in the Blue Nile River Basin. VU Amsterdam, 11 November 2013;

Tesfaye, A., & Brouwer, R. (2016). Exploring the scope for transboundary collaboration in the Blue Nile river basin: assessing downstream willingness to pay for upstream land use change to improve irrigation water supply. Environment and Development Economics, 21(2), 180-204. https://doi.org/10.1017/S1355770X15000182

Tignor, R.L. (1963) British agricultural and Hydraulic Policy in Egypt, 1882-1892. *Agricultural History* 37:2 63-74. 456

Tignor, R.L. (1966) *Modernization and British Colonial Rule in Egypt, 1882-1914*. Princeton: Princeton University Press,

Tignor, R. (1987) 'The Sudanese Private Sector. A Historical Overview', *The Journal of Modern African Studies* 25(2): 179–212.

Thiruvarudchelvan, T. (2010) Irrigation performance of Gezira scheme in Sudan: Assessment of irrigation efficiency using satellite data. *MSc Thesis UNESCO-IHE.*

Tomich, D. (1994). Small islands and huge comparisons: Caribbean plantations, historical unevenness, and capitalist modernity. Social Science History, 18(3), 339-357.

Tronvoll, K. (2009) War and the politics of identity in Ethiopia. Woodbridge, UK: James Currey.

Tsing, A.L. (2005) Friction, An ethnography of global connection, Princeton University Press

Turner (1893) The significance of the frontier in American history

Turner, R. and G. Daily (2008) The ecosystem services framework and natural capital conservation, Environmental and Resource Economics 39(1): 25-35.

Turnhout E. (2018) The Politics of Environmental Knowledge. Conservation Society ;16: 363-71

Tvedt, T. (2004) The River Nile in the Age of the British: Political Ecology and the Quest for Economic Power. London: I.B. Tauris.

UN Water (2013). Water Security and the Global Water Agenda. UN-Water Analytical Brief, UNU-INWEH

UN Water (2016). Integrated Monitoring Guide for SDG 6 Targets and global indicators, available online http://www.unwater.org/app/uploads/2017/03/SDG-6-targets-and-global-indicators_2016-07-19.pdf, accessed 08/08/2019

UNEP (2014) Nile Basin Adaptation to Water Stress Comprehensive Assessment of Flood & Drought Prone Areas

UNESCO-IHE, Addis Ababa University, IWMI, University of Khartoum, Vrije Universiteit Amsterdam, Delft University of Technology (2007) In Search of Sustainable Catchments and Basin-wide Solidarities; Transboundary Water Management of the Blue Nile River Basin, Research proposal submitted to NWO-WOTRO

UNESCO-IHE, IWMI, Hydraulics Research Center- Sudan, University of Khartoum, RISE American University in Cairo (2014) Inclusive Accounting for Nile waters: connecting investments in large scale irrigation to gendered reallocations of water and labor in the Eastern Nile basin, Project Proposal Submitted to CGIAR

UNICEF (2017) UNICEF Annual Report 2017 Egypt, accessed online https://www.unicef.org/about/annualreport/files/Egypt_2017_COAR.pdf on 1 August 2018

UNDP (United Nations Development Programme, United Nations Environment Programme), World Bank and World Resources Institute (2008) World Resources 2008: Roots of Resilience – Growing the Wealth of the Poor. Washington, DC: World Resources Institute.

USBR (1964) Bureau of Reclamation, United States Department of Interior. *Land and water resources of the Blue Nile Basin: Ethiopia. Main report and appendices I-V.* Washington D.C.: United States Government Printing Office.

Van der Zaag, P. (2003). The bench terrace between invention and intervention: Physical and political aspects of a conservation technology. In: A. Bolding, J. Mutimba, P. van der Zaag (eds.) *Agricultural intervention in Zimbabwe: New perspectives on extension* (184–205). Harare: University of Zimbabwe Publications.

Vatikiotis, P.J. (1991) *History of Modern Egypt: From Muhammad Ali to Mubarak, 4th Ed.* London: Weidenfeld and Nicolson,

Vaughan, S. (2011). Revolutionary Democratic State-building: Party, State and People in the EPRDF's Ethiopia. *Journal of Eastern African Studies* 5 (4) 619–640. doi:10.1080/17531055.2011.642520.

Veilleux, J. C. (2013) The Human Security Dimensions of Dam Development: The Grand Ethiopian Renaissance Dam. GLOBAL DIALOGUE, 15.

Verhoeven H. (2011) 'Dams are development': China, the Al-Ingaz regime and the political economy of the Sudanese Nile. In *Sudan looks east* (eds D Large, LA Patey), pp. 120–138. Oxford, UK: James Currey.

Verhoeven, H. (2011). Climate change, conflict and development in Sudan: Global neo-Malthusian narratives and local power struggles. Development and Change, 42, 679–707. doi:10.1111/j.1467-7660.2011.01707.x.

Verhoeven, H. (2015). Water, civilisation and power in Sudan. The political economy of Military-Islamist state-building. Cambridge: Cambridge University Press.

Vermillion, D. (2006) Lessons Learned and to be Learned about Irrigation Management Transfer, Paper presented at the Festschrift for E. Walter Coward, Jr., Ubud, Bali, 23 June 2006

Vitalis, R. (1995). *When Capitalists Collide: Business Conflict and the End of Empire in Egypt.* Berkeley, University of California Press.

Voll, S. P. (1980). "Egyptian Land Reclamation since the Revolution." Middle East Journal 34 (2):127–148.

Vörösmarty CJ et al. 2010 Global threats to human water security and river biodiversity. Nature 467, 555–561. (doi:10.1038/nature09440)

Wallach, B. (1988) Irrigation in Sudan Since Independence. *Geographical Review* 78(1) 417-434.

Warner, J.F. (2008) The politics of flood insecurity : framing contested river management projects, PhD thesis, Wageningen University

Warner, J.F. (2013). The Toshka mirage in the Egyptian desert – river diversion as political diversion. Environmental Science & Policy, 30, 102–112. doi:10.1016/j.envsci.2012.10.021.

Warner, J F., Wesselink A. J., Geldof G. D. (2018) The politics of adaptive climate management: Scientific recipes and lived reality. WIREs Climate Change. doi: 10.1002/wcc.515

Warner, J., Wester, P., Bolding, A. (2008) Going with the flow: river basins as the natural units for water management. Water Policy 10, 121–138. (doi:10.2166/wp.2008.210)

Waterbury, J. (1979). Hydropolitics of the Nile Valley. Syracuse University Press, Syracuse, NY.

Waterbury, J. and Whittington, D. (1998) Playing Chicken on the Nile? The Implications of Microdam Development in the Ethiopian Highlands and Egypt's New Valley Project. *Natural Resources Forum*, 22(3), 155-163.

Watts, M (2011) Silent Violence: Food, Famine and Peasantry in Northern Nigeria. 2nd Edition, Georgia Press

Watts, M. J. (2013) A Political Ecology of Environmental Security. In *Environmental Security: Approaches and Issues*, edited by Rita Floyd and Richard A. Matthew, 82–101. London: Routledge.

WCD (2000) World Commission on dams - Dams and Development: A New Framework for Decision-Making - The Report of the World Commission on Dams, Earthscan Publications

Wekker, G. (2016) White Innocence -Paradoxes of Colonialism and Race, Duke University Press,

Wellmon, C. (2010). Goethe's Morphology of Knowledge, or the Overgrowth of Nomenclature. Goethe Yearbook 17, 153-177.

Wesselink A, Kooij M, Warner J (2017) Socio-hydrology and hydrosocial analysis: towards dialogues across disciplines. *WIREs Water* 4(2)

Wester, P. (2008) Shedding the waters: institutional change and water control in the Lerma–Chapala basin, Mexico. PhD dissertation, Wageningen University, Wageningen, The Netherlands.

WFP (2019) World Food Programme Sudan Country Brief March 2019

Whatmore, S. (2006). Materialist returns: practising cultural geography in and for a more-than-human world. *Cultural Geographies* 13, 600-609

White, R. (1995) The Organic Machine – The remaking of the Columbia River, Hill and Wang, New York.

Whittington, D., and Guariso, G. (1983) *Water management models in practice: a case study of theAswan high dam*. Amsterdam: Elsevier.

Whittington, D., Waterbury, J., and Jeuland, M. (2014) The grand renaissance dam and prospects for cooperation on the  Eastern Nile Water Policy 16 595–608

Willcocks, W. (1894) *Report on Perennial Irrigation and Flood Protection for Egypt*. Cairo: National Printing Office,

Willcocks, W. (1904) *The Nile in 1904*, London: E.&F.N. Spon.

Willcocks, W. (1919) *Nile Projects*. Printing Office of the French Institute of Oriental Archaeology.

Willcocks, W. and J.I. Craig (1913) Egyptian Irrigation. 3rd Edition. London: E.&F.N. Spon.

Wittfogel, K. (1957) Oriental Despotism: A Comparative Study of Total Power. New Haven, Yale University Press

Woertz, E. (2013). Oil for Food: the Global Crisis and the Middle East. Oxford University Press, New York, NY, 319 pp.

Wolf, A.T. (1999) "Water Wars" and Water Reality: Conflict and Cooperation Along International Waterways. In: Lonergan S.C. (eds) Environmental Change, Adaptation, and Security. NATO ASI Series (2. Environment), vol 65. Springer, Dordrecht

Wolf A.T., ed. (2005). Hydropolitical Vulnerability and Resilience along International Waters. (Volume I: Africa). Nairobi: UN Environmental Program

Wolf, A.T. (2009) "A Long Term View of Water and International Security." *Journal of Contemporary Water Research & Education*. Issue 142, pp. 67-75

Wolf, A. T., Annika Kramer, Alexander Carius, and Geoffrey D. Dabelko (2005) "Managing Water Conflict and Cooperation." Chapter 5 in Worldwatch Institute. State of the World 2005: Redefining Global Security. Washington DC: Worldwatch Institute.

World Bank (1979) Arab Republic of Egypt: Agricultural development project (Minufyiyya – Sohag)

World Bank (1982) Sudan Incentives for Irrigated Cotton- Progress Towards Reform. World Bank Report No. 3810 SU

World Bank (1993) Water Resources Management A World Bank Policy paper

World Bank (1994) A Strategy for Managing Water in the Middle East and North Africa. Directions in Development, Washington, DC: World Bank.

World Bank (1994) Egypt - Irrigation Improvement Project (English). Washington, DC: World Bank. http://documents.worldbank.org/curated/en/518461468021582182/Egypt-Irrigation-Improvement-Project

World Bank (2000) Sudan - Options for the Sustainable Development of the Gezira Scheme, World Bank Report No. 20398-SU.

World Bank (2001) "Egypt: Toward Agricultural Competitiveness in the 21st Century, an Agricultural Export-Oriented Strategy." Report No. 23405, World Bank, Washington, DC.

World Bank (2007) Making the most of scarcity : Accountability for better water management results in the Middle East and North Africa (English). MENA development report. Washington, DC: World Bank.

World Bank (2009) Sudan – The road toward sustainable and broad-based growth, Poverty Reduction and Economic Management Unit, Africa Region, Washington DC

World Bank (2010) The World Bank and the Gezira Scheme in the Sudan – Political Economy of Irrigation Reforms, World Bank Report No. 69873

World Bank (2017) Ethiopia: Impacts of the Birr Devaluation on Inflation, Brief 121807, November 8, 2017

World Bank (2018) Water's Edge: Rising to the Challenge of a Changing World (English). Washington, D.C.:WorldBankGroup. http://documents.worldbank.org/curated/en/501621542313008673/Waters-Edge-Rising-to-the-Challenge-of-a-Changing-World

World Bank (2018a) New US$1 Billion Support to Egypt for Enabling Private Sector Led Job Creation, Press release 4 Dec 2018, https://www.worldbank.org/en/news/press-release/2018/12/04/new-us1-billion-support-to-egypt-for-enabling-private-sector-led-job-creation

Worster, D. (1985) Rivers of Empire: Water, Aridity, and the Growth of the American West. New York: Oxford University Press.

Yalew, S.G.; Mul, M.L.; Van Griensven, A.; Teferi, E.; Priess, J.; Schweitzer, C.; Van Der Zaag, P. (2016) Land-Use Change Modelling in the Upper Blue Nile Basin. Environments, 3, 21.

Yeh, E. (2012) Taming Tibet: Landscape Transformation and the Gift of Chinese Development, Cornell University Press, Ithaca

Yihdego, Z., Rieu-Clarke, A., & Cascão, A. E. (2016). How has the Grand Ethiopian Renaissance Dam changed the legal, political, economic and scientific dynamics in the Nile Basin? Editors' Introduction . Water International, 41(4).

Yoffe, S., Wolf AT, Giordano M. (2003) Conflict and cooperation over international freshwater resources: indicators of basins at risk. J. Am. Water Resour. Assoc. 39, 1109–1126. (doi:10.1111/j.1752-1688.2003.tb03696.x)

Yousif, G.M. (1997) The Gezira Scheme – The Greatest on Earth, Africa University House for Printing, Khartoum

Zegwaard, A. (2016) Mud - Deltas dealing with uncertainties, PhD thesis, Vrije Universiteit Amsterdam

Zeitoun M, Lankford B, Krueger T, Forsyth T, Carter R, Hoekstra AY, Taylor R, Varis O, Cleaver F, Boelens R, Swatuk L, Tickner D, Scott CA, Mirumachi N, Matthews N (2016) Reductionist and integrative research approaches to complex water security policy challenges. Glob Environ Change 39:143–154. doi:10.1016/j. gloenvcha.2016.04.010

Zenawi, M. (2011) Hydro-power for Sustainable Development, Address by H.E. Mr. Meles Zenawi, Prime Minister of the Federal Democratic Republic of Ethiopia, 31 March 2011, Addis Ababa, Ethiopia

Zenawi, M. (2011a) The speech delivered by H.E. Prime Minister Meles Zenawi to mark the official commencement of the Millennium Dam project, April 02/2011, Guba, Beneshangul Gumuz, available online on https://hornaffairs.com/2011/04/02/ethiopia-great-dam-on-nile-launched/

Zhu, Z., Molden, D., El Kady, M., (1996). Salt Loading in the Nile Irrigation System. WRSR Discussion Paper Series No 25.

Zwarteveen, M., Kemerink-Seyoum, J.S., Kooy, M., Evers, J., Guerrero, T.A., Batubara, B., Biza, A.M., Boakye-Ansah, A.S., Faber, S., Flamini, A.C., Cuadrado-Quesada, G., Fantini, E., Gupta, J., Hasan, S., Horst, R., Jamali, H., Jaspers, F.G., Obani, P., Schwartz, K., Shubber, Z., Smit, H.V., Torío, P., Tutusaus, M., & Wesselink, A.J. (2017). Engaging with the politics of water governance. Wiley Interdisciplinary Reviews: Water. https://doi.org/10.1002/wat2.1245

Zwarteveen, M., Smit, H., Domínguez Guzmán, C., Fantini, E., Rap, E, Van der Zaag, P.and Boelens, R. (2018) Accounting for Water: Questions of Environmental Representation in a Nonmodern World. Chapter 11 in "Rethinking Environmentalism: Linking Justice, Sustainability, and Diversity, Strüngmann Forum Reports, vol. 23, series editor Julia Lupp. Cambridge, MA: MIT Press

# Summary

This dissertation examines Nile water security through the morphology of the river. Water projects are often legitimized by arguing that they will increase the reliability of water or increase its availability for abstract populations. Such analyses often leave unexplained who specifically benefit from these projects and, more so, who do not. Examining the morphology of the river – its form and structure – allows for a historical and material understanding of how hydraulic infrastructure and discourses of water security develop and what this means to whom. My aim is to better understand how scientists, engineers and water users engage in rearranging the morphology of the Nile and in so doing shape their relative positions vis-à-vis each other and the river. In this way the dissertation seeks to support more equitable and sustainable forms of Nile development.

The research is structured around three research questions:

1. Who and what make the modern limits to water security on the Nile and why so?
2. How do projects in the name of water security on the Nile transform the river and who benefits?
3. What does this imply for the use of science for understanding and changing the emerging patterns of Nile water distribution?

The sediments of the Nile literally ground my analysis of water security. A focus on practices through which the river bed is rearranged yields an understanding of how technologies and discourses of water security shape the river, and conversely, how these technologies and discourses are used as instruments to forge new patterns of Nile water distribution. Here, I build on two bodies of literature that partly overlap. First, I am inspired by a broad group of biologists, anthropologists, political scientists and philosophers of science who critique the modern scientific separation between technology and politics (Goethe 1987 [1776-1832], Foucault 2007 [1978], Haraway 1991, Latour 1993, Mitchell 2002). Second, I build on scholars of 'historical materialism' who analyse how highly uneven forms of governance emerge through rearrangements of the environment (Smith 1984, McMichael 2012, Moore 2015). Bringing into conversation insights from these two groups of scholars, I analyse how new forms of collaboration and valuation of water emerge with the modern development of the river.

I operationalize the research by bringing into relation a historical account of dam development on the Nile and three contemporary projects of river development. Taking the sediments of the Nile as the starting point enables a relational analysis of water security, one that acknowledges that rivers are a product of earlier relations deposited in the river bed and that also appreciates the possibilities of millions along the Nile working continuously to remake these relations. The morphological account of Nile river development thus created complements other accounts in three ways.

First, the analysis of Chapter 2 shows how three waves of investment in large scale hydraulic infrastructures over the last two hundred years have each shifted the definition of, and limits to, water security along the river. The first wave of construction of large barrages and dams – starting in the 19th century to enable the expansion of cotton cultivation – led to the depletion of almost the entire base flow of the Nile. The concerns that arose in regards to safeguarding this base flow from upstream consumption invoked the invention and conquest of the *Nile Basin* territory. A second wave of hydraulic construction took place from the 1950s to the 1970s. The postcolonial dams that were constructed on the upstream national boundaries of Egypt and Sudan enabled the storage of the peak flows of the Nile in the name of *national* development. With this, the definition of water security and the consumption of Nile waters came to encompass (almost) the entire river flow. Yet this did not stop governments and companies from investing in the contemporary, third, wave of mega-projects on the Nile, which started at the end of the 20th century. Faced with depleted rivers and old irrigation schemes that are no longer producing export commodities, international organisations such as the World Bank, the IMF, FAO and IWMI worked with governments and investors along the Nile to redefine water security in terms of water productivity. This redefinition – based on the simple idea that increasing the production of crops per drop is critical now that river waters have been depleted and global food requirements are rising – continues to be used by governments and investors as a plea to reallocate water to new 'more efficient' mega-projects.

Second, the accounts of drainage, irrigation reform and hydraulic reconstruction in Chapters 3, 4 and 5 respectively, demonstrate that the modern securitization of the river is not a story of limitless exploitation and domination over land, water and labour. On the contrary, with the rising environmental and political risks of diverting water to new hypermodern water projects, the prospects for such projects are reducing. The targeted subjects of modern river development make use of the new spaces for 'river development' thus created by carving out their own projects. They creatively mobilize old irrigation and drainage infrastructures in ways that escape the universal logic of modern water security.

Chapter 3 focuses on how the power of the Ethiopian state is inscribed in a hillslope of the Ethiopian highlands through what is arguably the Nile's largest modern hydraulic project in terms of labour: the state soil conservation programme. More than ten million people are mobilized each year to work on soil conservation for twenty to forty days. The chapter explores how access to drainage on a hillslope in the Choke Mountains is shaped by the engagements of its users in state conservation projects. This brings into view how the work of a generation of middle-aged male farmers on state-induced soil conservation both cements state power and enables the diversion of harmful drainage flows to lands held by elderly people that are sharecropped in by young landless families. As a result of this, soil erosion is not halted but aggravated. Moreover, landless families that fail to live up to the model of the 'farmer interested in soil conservation' created a competing 'trader model' with their own institutions of financing, oxen-sharing and a system of transport and barter with lowland markets. The denial of their 'trader identity' by landholders and officials fuels generational conflicts over drainage which deepen the fractures in the hill and pose new challenges to government authority. In this way, land degradation shapes both the powers and the limits of the 'developmental state'.

Chapter 4 analyses practices of planting of crops, irrigation and desilting of canals through which officials, tenants, sharecroppers and sediments enact the categories of Irrigation Management Transfer in the Gezira Irrigation Scheme in Sudan. A close examination of the changing cross section of a century old earthen irrigation canal during the 2011 irrigation season shows how the shape of the canal is influenced by, and influences, struggles over water and power. An analysis of the flows of sediment, water and crops along the canal, that supplies water to an area of 800 ha, shows how a network of male tenant investors took control over the newly established Water User Association and with it the scheme's agricultural service provision and canal maintenance. Its interest in the amount of cubic meters silt excavated – rather than the shape of the canal – resulted in large differences in water access along the canal. The plots with good access to water are largely under control of investors who employ male migrant labourers to grow high value cotton and vegetables. Most of the plots with unreliable access to water are sharecropped by migrant women. These women – numbering several hundreds of thousands in the Gezira but hardly featuring in official reports – deal with large uncertainties in access to water by enlarging the number of plots cultivated with drought-resistant sorghum and focusing on those plots for which they manage to secure access to water. By using the increasing amount of water diverted to the scheme – in 2011 some 10% of all Nile water – to cultivate sorghum and groundnuts, they largely contributed to the scheme producing more crops than ever before, providing grain, vegetables and vegetable oil for an estimated 4 million people.

Chapter 5 reconstructs the making of satellite-based 'water accounts' of water productivity as a historical and material process so as to shed a different light on both the potential and limitations of 'water accounting' for re-organising Nile waters. It analyses how practices of 'water accounting' support particular patterns of water distribution in, and around, an irrigation scheme near Khartoum. Brining into conversation accounts of 'water accounting' scientists with accounts of water users, the chapter reconstructs how racial and gender relations of crop cultivation were (re)produced with superimpositions of three irrigation grids since the 1970s. This analysis helps to explain why the recent efforts to 'upscale the success' of a high-tech private centre-pivot irrigation scheme, which produces alfalfa for export to Saudi Arabia, are not unfolding in ways suggested on the basis of water accounting maps or anticipated by the company. Faced with rising opposition from neighbouring smallholder farmers, who suffered from earlier expansions of the centre-pivot farm and rising food prices, the modern company fenced and consolidated its boundaries. Around the company farm, new niches of sorghum, vegetable and wheat farming are emerging. Interestingly, some of these initiatives rely on a steady supply of water by the company, which provides the water to the tail ends of its canals at water rates that are amongst the lowest in Sudanese irrigation. Because biomass growth on these plots is relatively low (especially during the cloudless part of the year for which remote sensing images can be processed) these new initiatives appear as 'unproductive' on the remotely sensed water accounting maps. By linking the making of such maps to those people and projects with an interest in these maps, the chapter brings to the fore the political nature of water accounting.

Third, Chapters 2 to 5 intend to refocus water security and accountability away from universal terms of resilience and good governance, towards historical and material accounts that highlight how frameworks of water security materialize in real diversions of water and sediment. Grounding the analysis in particular positions of the changing river morphology yields a limited and therefore partial account of river basin development. Rather than viewing this partiality as a disadvantage, the particular limits function as a spotlight that can actually enable holding people responsible for projects carried out in the name of water security (cf. Haraway 1991). The three projects show how the universal guidelines for participatory soil conservation, irrigation management and water accounting for water productivity fit particularly well with new liberal models of green inclusive development, but have little affinity with some of the newly emerging modes of collaboration over water along the Nile. Reflexivity and self-learning – increasingly popular recipes suggested by academics to peers and water users facing declining supplies of water – will by themselves not contribute to strengthening these emerging alternatives for more sustainable and just distributions of Nile waters. Only once we move away from the idea of science as a universal tool for discovery of the unknown, and move instead towards science as a mode of conversation about what sustainability and equity imply for whom, will it become clearer how scientific tools can effectively be mobilized to push for more emancipatory agendas. By creating this morphological account of Nile river development, and discussing it in classrooms, offices and the Senaatszaal of TU Delft, I hope to enlarge the space for some of the promising emerging patterns of water security that often remain hidden in accounts of academics, engineers and policy makers.

# SAMENVATTING

Dit proefschrift onderzoekt zekerheid van Nijlwater door te kijken naar de veranderende morfologie van de rivier. Veel moderne waterprojecten langs de Nijl claimen dat ze door water op te slaan, of door de waterproductiviteit te verhogen, bijdragen aan het vergroten van waterzekerheid langs de Nijl. Maar vaak blijft buiten beeld wat hier precies mee bedoeld wordt, wie er profiteert, en wie niet. Door te beginnen bij de morfologie van de rivier – haar vorm en structuur – brengt deze thesis in beeld hoe waterzekerheid vorm krijgt in waterinfrastructuur, organisaties en kennis. Mijn doel is om zo beter te begrijpen hoe wetenschappers, ingenieurs en watergebruikers bijdragen aan het herverdelen van de rivier en hoe dit hun onderlinge verhoudingen (her)bepaalt. De inzichten die dit oplevert kunnen gebruikt worden in nieuwe projecten die gericht zijn op een eerlijkere en duurzamere ontwikkeling van de Nijl.

Dit onderzoek beantwoordt drie onderzoeksvragen:

1.  Wie en wat bepalen wat zekerheid van Nijlwater is?

2.  Hoe veranderen moderniseringsprojecten de rivier en wie hebben hier voordeel van?

3.  Wat zijn hiervan de implicaties voor het gebruik van water kennis en wetenschap voor het begrijpen van de herverdeling van Nijlwater?

De sedimenten van de Nijl vormen - letterlijk - mijn analyse van water zekerheid. Mijn aandacht voor de praktijk waarin rivierbedding constant verandert, leert me hoe theorieën en technologieën van water zekerheid de rivier vormgeven, en omgekeerd, hoe deze technologieën en theorieën als instrumenten worden gebruikt om de waterverdeling van de Nijl te veranderen. Hierbij bouw ik voort op twee gedeeltelijk overlappende stromingen in de wetenschappelijke literatuur. Ten eerste, gebruik ik het werk van een brede groep van biologen, antropologen, politieke wetenschappers en wetenschapsfilosofen die laten zien hoe technologie en politiek altijd verweven zijn (Goethe 1987 [1776-1832], Foucault 2007 [1978], Haraway 1991, Latour 1993, Mitchell 2002). Ten tweede, ben ik geïnspireerd door wetenschappers die ecologische relaties een duidelijke plaats geven in de analyse van ongelijke geografische ontwikkeling (Smith 1984, McMichael 2012, Moore 2015). Samen helpen deze werken me te begrijpen hoe de moderne ontwikkeling van de rivier vorm geeft aan nieuwe waarden en manieren van samenwerking bij het verdelen van Nijlwater.

Ik operationaliseer mijn aanpak door een historisch analyse van het bouwen van dammen op de Nijl in verband te brengen met drie eigentijdse projecten van rivierontwikkeling. Door de sedimenten van de Nijl als startpunt te nemen wordt een relationele definitie van waterzekerheid mogelijk: zo'n definitie doet zowel recht aan de historische relaties die de rivier vormden als aan de verschillende mogelijkheden van miljoenen mensen die dagelijks werken om deze relaties te hervormen. De morfologische kijk op de ontwikkeling van de Nijl die dit oplevert draagt op drie manieren bij aan bestaande inzichten.

Ten eerste, laat hoofdstuk 2 zien hoe drie investeringsgolven in grootschalige hydraulische infrastructuur op de Nijl in de laatste 200 jaar gepaard gingen met veranderende definities en grenzen van waterzekerheid. De eerste golf van investeringen in grootschalige hydraulische werken, die begon in de 19e eeuw om uitbreiding van de katoenteelt mogelijk te maken, leidde tot het gebruik van vrijwel de complete permanente afvoer van de Nijl. Bovendien droegen zorgen over aanvoer van dit water bij aan de vorming van het begrip 'Nijl stroomgebied' en de verovering hiervan. Een tweede golf van grote investeringen in waterprojecten vond plaats van de jaren 1950 tot de jaren 1970. De postkoloniale mega-dammen die werden gebouwd op de stroomopwaartse grenzen van Egypte en Soedan maakten de berging van de piekafvoer van de Nijl mogelijk voor *nationale* ontwikkeling. Zowel het begrip van waterzekerheid en het gebruik van water werd hiermee verruimd naar (vrijwel) de totale rivierafvoer. Dit weerhield overheden en bedrijven er niet van om te investeren in de huidige, derde, golf van mega-Nijlprojecten. Geconfronteerd met rivieren waarvan vrijwel al het water werd geconsumeerd door steeds slechter renderende waterprojecten in landen met hoge schuldenlasten, herdefinieerde onder andere de Wereldbank, het Internationaal Monetair Fonds, de Voedsel- en Landbouworganisatie van de Verenigde Naties en het International Water Management Instituut (IWMI) de definitie van waterzekerheid in termen van waterproductiviteit. Deze hernieuwde definitie – gestoeld op het simpele idee dat nu al het water gebruikt wordt en de vraag naar voedsel toeneemt, het noodzakelijk is om de gewasproductie per druppel water te verhogen – wordt vandaag de dag volop gebuikt om het afleiden van Nijlwater naar nieuwe megaprojecten te legitimeren.

Ten tweede, laten de analyses van drainage, de hervorming van irrigatie en de constructie van een nieuw irrigatiesysteem in hoofdstuk 3, 4, en 5 zien dat de veranderende ideeën en projecten in naam van waterzekerheid niet te vatten zijn in woorden van uitbuiting en controle over land, water en arbeid. Integendeel, nu de ecologische en politieke kosten van nieuwe megaprojecten steeds hoger en breder zichtbaar worden, nemen de mogelijkheden tot het gemakkelijk controleren van water en het behalen van grote winsten zienderogen af. Dit geeft ruimte aan mensen die achtereenvolgens werden gelabeld als achterlijk, onderontwikkeld en niet veerkrachtig om hun eigen projecten van waterzekerheid vorm te geven. Zij mobiliseren oude irrigatie- en drainagewerken op wijzen die niet te vatten zijn in 'moderne' termen van waterzekerheid (zoals waterproductiviteit en individuele eigendomsrechten).

Hoofdstuk 3 richt zich op de vraag hoe de macht van de Ethiopische staat vorm krijgt in een berghelling, door wat waarschijnlijk het grootste hedendaagse Nijlproject is in termen van arbeid: het staats-bodembeschermingsprogramma. Meer dan 10 miljoen mensen worden elk jaar voor 20 tot 40 dagen gemobiliseerd om te werken aan erosie beperkende maatregelen. Dit hoofdstuk verkent hoe toegang tot drainage op de helling wordt gevormd door de landgebruikers die deelnemen aan deze projecten. Het laat zien hoe een generatie van boeren van middelbare leeftijd door het project zowel de macht van de staat gestalte geeft als overvloedig drainagewater naar door jonge landloze families gepachte kavels leidt. De erosie van de bodem neemt hierdoor niet af maar juist toe. Onder landloze families die niet aan het beeld van 'de boer geïnteresseerd in bodembescherming' kunnen voldoen leeft een alternatief rolmodel van 'de handelaar', met eigen instituties van financiering, het delen van lastdieren en handel met markten in het laagland. De consequente ontkenning van deze 'handelaarsidentiteit' door ambtenaren en landeigenaren leidt tot een verergering van generatieconflicten over onder andere drainage en een groeiend verzet tegen de overheid die iedereen als boeren met land blijft behandelen. Hiermee vormt landdegradatie zowel de macht als de grenzen van het Ethiopische staatsapparaat.

Hoofdstuk 4 analyseert de praktijk van gewaskeuze, irrigatie en het onderhoud van aarden kanalen waardoor het beleid van Irrigatie Management Transfer vorm krijgt in het Gezira irrigatieschema in Sudan. Door de veranderende dwarsdoorsnede van een 100 jaar oud aarden irrigatiekanaal onder de loep te nemen tijdens het irrigatieseizoen van 2011, laat dit hoofdstuk zien hoe het kanaal vorm geeft aan en vorm krijgt door de veranderende verdeling van water en macht. Een analyse van stromen van water, sediment en landbouwgoederen langs het kanaal, dat water levert aan een gebied van 800 ha, laat zien hoe een machtig netwerk van investeerders de controle over de nieuw opgerichte Water User Association grijpt en daarmee ook de agrarische dienstverlening en het kanaalonderhoud onder controle krijgt. Doordat er voor het kanaalonderhoud afgerekend wordt per uitgegraven kubieke meters slib – en de vorm van het kanaal daarom minder interessant is voor de aannemers die worden ingeschakeld om de kanalen uit te graven – ontstaan grote verschillen in de kanaaldoorsnedes en daarmee grote verschillen in de toegang tot water. De percelen met goede toegang tot water zijn grotendeels onder controle van de investeerders die mannelijke arbeidsmigranten in dienst hebben om katoen en groenten te verbouwen. De percelen met een onbetrouwbare watervoorziening worden veelal gedeelpacht door migrantenvrouwen. Deze migrantenvrouwen – waarvan er honderdduizenden zijn in de Gezira, maar die nauwelijks genoemd worden in officiële rapporten - hebben verschillende manieren om met de onzekere toegang tot water om te gaan. Ze verbouwen een groot deel van de percelen onder hun hoede met droogtebestendige sorghum. Daarnaast verbouwen ze vaak meerdere percelen tegelijk en richten ze zich op die velden waar ze erin slagen om in de juiste periode te irrigeren. Door de toenemende hoeveelheid water die van de Blauwe Nijl naar het Gezira irrigatiesysteem wordt afgeleid (in 2011 ongeveer 10% van al het Nijlwater) te gebruiken om sorghum en pinda's te verbouwen, droegen deze vrouwen in belangrijke mate bij aan de recordopbrengsten die in de jaren van dit onderzoek in het irrigatiesysteem geboekt werden. Het voedsel dat geproduceerd wordt in de Gezira is naar schatting goed voor het voorzien van graan, groenten en plantaardige olie voor zo'n 4 miljoen mensen.

Hoofdstuk 5 reconstrueert het maken van 'water accounts', die op basis van satellietbeelden de waterproductiviteit in beeld brengen, als een historisch en materieel proces. Op deze wijze worden zowel het potentieel als de beperkingen van 'water accounting' voor het reorganiseren van Nijlwater nader in beeld gebracht. Het hoofdstuk brengt naar voren hoe door het maken van 'water accounts' de wenselijkheid van bepaalde vormen van watergebruik in en rond een irrigatieschema in de buurt van Khartoem wordt gesuggereerd. Door de op basis van remote sensing gemaakte water accounts in verband te brengen met de 'accounts' van watergebruikers rond het irrigatieschema, reconstrueert het hoofdstuk hoe raciale en gender relaties van landbouw in het gebied ge(re)produceerd werden door de bouw van drie over overlappende irrigatiesystemen sinds de jaren 1970. De analyse helpt te begrijpen waarom recente pogingen om het 'succes op te schalen' van het nieuwste high-tech center pivot irrigatiesysteem, dat in de afgelopen 10 jaar werd gebouwd voor de productie van veevoer voor export naar Saudi Arabië, allemaal strandden. Geconfronteerd met de toenemende weerstand van naburige kleine boeren die land verloren door eerdere uitbreidingen en die voedselprijzen sterk zagen stijgen, heeft de eigenaar van het pivot-irrigatie systeem zich gericht op de consolidatie van haar grenzen in plaats van de uitbreiding daarvan. Rond de high-tech pivot-irrigatie boerderij ontstaan nieuwe niches van sorghum-, groente- en tarweteelt. Interessant is dat sommige van deze initiatieven van water voorzien worden door de kanalen van het pivot-irrigatie bedrijf, dat het water aanbiedt tegen de laagste tarieven voor irrigatiewater in Soedan. Omdat de biomassagroei op deze percelen relatief laag is (vooral tijdens het onbewolkte deel van het jaar waarvoor remote sensing beelden kunnen worden verwerkt), verschijnen deze nieuwe initiatieven als 'niet-productief' op waterproductiviteitskaarten. Door de productie van dergelijke kaarten in verband te brengen met de mensen en projecten die een belang hebben in het land en water op deze kaarten, komt het politieke karakter van 'water accounting' scherp in beeld.

Ten derde, wordt met hoofdstukken 2 tot en met 5 een shift gemaakt van het begrijpen van waterzekerheid in universele termen van veerkracht en goed bestuur, naar een historische en materieel begrip van waterzekerheid als een product van samenwerkingen en strijd over water, sediment en kennis. Door de analyse te verankeren in specifieke posities in de verdeling van Nijl water, wordt een gelimiteerd beeld van de ontwikkeling van de Nijl geconstrueerd. Dit is echter niet slechts een beperking, maar helpt om focus te creëren en daardoor mensen verantwoordelijk te kunnen houden (cf. Haraway 1991). De drie projecten laten zien hoe de universele recepten voor participatieve bodemscherming, de transfer van irrigatiebeheer naar gebruikers en het maken van waterproductiviteitskaarten bijzonder goed aansluiten bij nieuwe liberale modellen van groene inclusieve ontwikkeling, maar weinig affiniteit hebben met nieuwe vormen van samenwerking in de projectgebieden. Reflexiviteit en zelfreflectie – de nieuwe academische buzzwoorden die zowel aan academici als watergebruikers worden opgedrongen - zullen op zichzelf niet bijdragen aan het versterken van deze opkomende alternatieven voor duurzamere en rechtvaardigere herverdelingen van Nijlwater. Alleen wanneer we afstappen van het idee van wetenschap als een universeel hulpmiddel voor het ontdekken van het onbekende, en in plaats daarvan wetenschap gaan zien als een gespreksvorm over wat duurzaamheid en rechtvaardigheid betekent en voor wie, zal het duidelijker worden hoe wetenschap kan bijdragen aan meer emancipatoire agenda's. Door het schrijven van deze morfologische analyse van de Nijl ontwikkeling, en het bespreken ervan in klaslokalen, kantoren en in de Senaatszaal van de TU Delft, hoop ik bij te dragen aan het succes van het enkele veelbelovende patronen van waterzekerheid langs de Nijl die vaak verborgen blijven in schriften van academici, ingenieurs en beleidsmakers.

# ABOUT THE AUTHOR

Hermen Smit teaches Water Governance at IHE Delft Institute for Water Education. His work focuses on the design, construction and manipulation of water infrastructures to understand the politics of water engineering.

After receiving his MSc in civil engineering with distinction from TU-Delft in 2003, he worked as a consultant for Water Boards in the Netherlands and in Bangladesh on the assessment, operation and maintenance of drainage systems. In 2008, Hermen joined UNESCO-IHE (now IHE Delft) as a PhD fellow and (later) assistant professor in Water Governance. Hermen's PhD research focused on the politics of water development in the Nile basin (with a focus on Ethiopia and Sudan). He currently teaches the courses 'who and what make water management expertise?' and 'institutional analysis'.

Since 2008, Hermen has been involved as a researcher, advisor and project manager in several water research and education development projects. Between 2012 and 2016, he worked on a project with 8 Ethiopian universities to make irrigation education in Ethiopia more practice oriented and interdisciplinary. Between 2014 and 2017, he contributed to the Accounting for Nile Waters research project about the politics of representing water (see www.nilewaterlab.org). In 2019, Hermen started working on an education and research project about 'the material politics of community involvement in climate proofing Rotterdam' and on a project with the Ministry of Water Resources and Irrigation in post-revolution Sudan to train staff and set up new research about water management to support Sudanese food producers.

Netherlands Research School for the
Socio-Economic and Natural Sciences of the Environment

# D I P L O M A

## For specialised PhD training

The Netherlands Research School for the
Socio-Economic and Natural Sciences of the Environment
(SENSE) declares that

# Hermen Smit

born on 18 July 1979 in Amsterdam, The Netherlands

has successfully fulfilled all requirements of the
Educational Programme of SENSE.

Delft, 17 December 2019

The Chairman of the SENSE board

Prof. dr. Martin Wassen

the SENSE Director of Education

Dr. Ad van Dommelen

The SENSE Research School has been accredited by the Royal Netherlands Academy of Arts and Sciences (KNAW)

K O N I N K L I J K E   N E D E R L A N D S E
A K A D E M I E   V A N   W E T E N S C H A P P E N

The SENSE Research School declares that Hermen Smit has successfully fulfilled all requirements of the Educational PhD Programme of SENSE with a work load of 33.6 EC, including the following activities:

### SENSE PhD Courses

o   Techniques for Writing and Presenting Scientific Papers (2009)
o   Environmental research in context (2010)
o   Research in context activity: 'Initiating and co-organizing Nile water lab website launch and mini-symposium conversations between remote sensing experts, water engineers and social scientists on: 'Accounting for Nile waters' (IHE Delft, 24-25 November 2016)"

### Other PhD and Advanced MSc Courses

o   Storytelling, IHE Delft (2018)

### Management skills Training and Activities

o   Member of organising committee "New Nile Perspectives" conference in Khartoum, Sudan, 6-8 May 2013
o   Programme coordinator Water Management MSc programme UNESCO-IHE (2013-2015)

### Didactic Skills Training

o   Supervising MSc thesis of nine MSc students (2010-2017)
o   Teaching in the MSc courses 'Programme introduction/Principles of Integrated Water Resources Management', 'The water resources system', 'Institutional analysis', Research methodology' and IHE's International field trip (2010-2018)

### Selection of Oral Presentations

o   *Negotiating environmental change - The creation of a gully in the Ethiopian highlands.* Workshop on Capturing Critical Institutionalism, Conference of the American Association of Geographers, 11-14 April 2011, Seattle, United States of America
o   *River Sclerosis – The political morphology of damming the Eastern Nile basin 1900-2015.* International Conference of the European Network of Political Ecology (ENTITLE), 20-24 March 2016, Stockholm, Sweden
o   *Eastern Nile developments: (re)shaping water and sediment flows in the Gezira Irrigation Scheme - Sudan.* New Nile Opportunities: Scientific advances towards Prosperity in the Eastern Nile Basin, ENTRO, 8-9 December 2014, Addis Ababa, Ethiopia
o   *From Bounded agency to actualizing history- shaping canals, land tenure and citizenship in the Gezira Scheme, Sudan.* Conference of the American Association of Geographers, 5-8 April 2017, Boston, United States of America

SENSE Coordinator PhD Education

Dr. ir. Peter Vermeulen